Advances in Geographical and Environmental Sciences

Series Editor
Dr. R.B. Singh

More information about this series at http://www.springer.com/series/13113

AIMS AND SCOPE

Advances in Geographical and Environmental Sciences synthesizes series diagnostigation and prognostication of earth environment, incorporating challenging interactive areas within ecological envelope of geosphere, biosphere, hydrosphere, atmosphere and cryosphere. It deals with land use land cover change (LUCC), urbanization, energy flux, land-ocean fluxes, climate, food security, ecohydrology, biodiversity, natural hazards and disasters, human health and their mutual interaction and feedback mechanism in order to contribute towards sustainable future. The geosciences methods range from traditional field techniques and conventional data collection, use of remote sensing and geographical information system, computer aided technique to advance geostatistical and dynamic modeling.

The series integrate past, present and future of geospheric attributes incorporating biophysical and human dimensions in spatio-temporal perspectives. The geosciences, encompassing land-ocean-atmosphere interaction is considered as a vital component in the context of environmental issues, especially in observation and prediction of air and water pollution, global warming and urban heat islands. It is important to communicate the advances in geosciences to increase resilience of society through capacity building for mitigating the impact of natural hazards and disasters. Sustainability of human society depends strongly on the earth environment, and thus the development of geosciences is critical for a better understanding of our living environment, and its sustainable development.

Geoscience also has the responsibility to not confine itself to addressing current problems but it is also developing a framework to address future issues. In order to build a 'Future Earth Model' for understanding and predicting the functioning of the whole climatic system, collaboration of experts in the traditional earth disciplines as well as in ecology, information technology, instrumentation and complex system is essential, through initiatives from human geoscientists. Thus human geosceince is emerging as key policy science for contributing towards sustainability/survivality science together with future earth initiative.

Advances in Geographical and Environmental Sciences series publishes books that contain novel approaches in tackling issues of human geoscience in its broadest sense – books in the series should focus on true progress in a particular area or region. The series includes monographs and edited volumes without any limitations in the page numbers.

Osvaldo Muñiz Solari • Ali Demirci
Joop van der Schee
Editors

Geospatial Technologies and Geography Education in a Changing World

Geospatial Practices and Lessons Learned

Springer

Editors
Osvaldo Muñiz Solari
Department of Geography
Texas State University
San Marcos, TX, USA

Ali Demirci
Department of Geography
Fatih University
Istanbul, Turkey

Joop van der Schee
Faculty of Geosciences
Utrecht University
Utrecht, The Netherlands

Faculty of Earth and Life Sciences
VU University Amsterdam
Amsterdam, The Netherlands

ISSN 2198-3542 ISSN 2198-3550 (electronic)
Advances in Geographical and Environmental Sciences
ISBN 978-4-431-55518-6 ISBN 978-4-431-55519-3 (eBook)
DOI 10.1007/978-4-431-55519-3

Library of Congress Control Number: 2015949261

Springer Tokyo Heidelberg New York Dordrecht London

Printed on acid-free paper

Springer Japan KK is part of Springer Science+Business Media (www.springer.com)

Preface

Within a few decades, computers and the Internet really have changed the world. Although education is not in the front line of this revolution, it cannot lag behind, as modern education should prepare students for the world of today and tomorrow. For geography education the digital revolution offers many opportunities. Both formal and informal expressions of learning are being influenced by geospatial technologies. These technologies have the potential to enhance students' twenty-first-century skills and can stimulate a new way of learning or at least offer better opportunities to develop higher-order thinking skills. Modern geospatial technologies can help learners to gain a better view of the world and provide opportunities for learners to better understand the planet Earth and attempt to resolve geographical issues.

This book gives an overview of the state of the art in the field where geospatial technologies and geography education meet. Geospatial practices are presented as an important means of learning about and with these technologies. The 17 chapters of the book are organized in five parts that deal with a theoretical base, implementation of geospatial technologies in formal and informal education, the training of teachers, evaluation and assessment, and recommendations for the near future.

Geography today cannot study geographical issues and resolve them without an effective use of geospatial technologies. That is why geography education must incorporate their use and applications in formal education and informal learning about the Earth's phenomena. The Commission on Geographical Education of the International Geographical Union wants to reach a wide spectrum of people interested not only in geospatial technologies but also in geography education. For this purpose an e-book is also available to allow people from different parts of the world to become informed about our analyses, propositions, and recommendations. Print copies of this book certainly can reach the libraries and centers of study in several countries; however, we want to be sure that this book also reaches people, institutions, and countries where scientific contributions are more difficult to obtain through the traditional way of publishing. In order to complete this job, we encourage people around the world to think of possible ways to transfer this body of knowledge to their communities in different languages. Then our effort will be a total success, because we want to offer this book to all people.

Texas, TX, USA
Istanbul, Turkey
Utrecht, The Netherlands
Amsterdam, The Netherlands

Osvaldo Muñiz Solari
Ali Demirci
Joop van der Schee

Acknowledgments

The editors would like to recognize the International Geographical Union Commission on Geographical Education (IGU-CGE) that encouraged some members to take the role of editors and guide the creation of this book.

The editors extend their thanks to Prof. R.B. Singh, vice president of IGU, for his assistance in facilitating the initial process of submitting the proposal for this book to Springer.

The editorial guidance by Ms. Taeko Sato in Springer Japan's Editorial Department is also acknowledged. Her assistance was very valuable as she provided advice and suggestions during the preparation of the manuscripts.

The authors of the chapters were always available and showed great willingness to develop the manuscripts and respond to the review process very efficiently. The editors appreciated their commitment to this project, which we believe will be a good contribution to geography education.

Contents

Abbreviations

AGO	ArcGIS online
AGSI	African Geospatial Sciences Institute
BISEL	Biotec Index at secondary education level
BYOD	Bring your own device
CBI	Content-based information
CGIS-NUR	Center for GIS and Remote Sensing – National University of Rwanda
CIM	Collaborative inquiry model
CK	Content knowledge
COPPA	Children's Online Privacy Protection
CPs	Communities of practice
CTE	Career and technology education
DE	Digital earth
ELLs	English language learners
ESSIP	Earth System Science Internet Project, Wyoming, USA
FCAT	Florida Comprehensive Achievement Test
FERPA	Family Educational Rights and Privacy Act
GCSE	General Certificate of Secondary Education, UK
GE	Geography education
Geo-ITC	Geoinformation and Earth observation
GI	Geoinformation
GIS	Geographic information science
GPS	Global Positioning System
GPs	Geospatial practices
GSTs	Geospatial technologies
GTCM	Geospatial technology competency model
IBL	Inquiry-based learning
ICT	Information and communication technology
iGETT	Integrated geospatial education technology and training
IL	Informal learning

IT	Information technology
IYGU	International Year of Global Understanding, UN
LMC	Learning cluster model
MI	Multiple intelligences
MRI	Magnetic resonance imaging
MOOCs	Massively open online courses
NCGE	National Council for Geographic Education
NRC	National Research Council
NSDI	National Spatial Data Infrastructure
OSM	OpenStreetMap
PBL	Problem-based learning, also project-based learning
PCK	Geographic-didactic knowledge
PK	Pedagogical knowledge
PPGIS	Public participatory geographic information system
RS	Remote sensing
SaaS	Software-as-a-service
SPACIT	Spatial citizenship
SOA	Service-oriented architecture
STAT	Spatial thinking abilities test
STEM	Science, technology, engineering, and math
TCK	GIS-knowledge
TERC	Teacher Educational Resource Center
TK	Technological knowledge
TPACK	Technological pedagogical content knowledge, originally TPCK
TPCK	(equivalent to GIS-didactic knowledge)
TPK	Technological pedagogical knowledge
TTF	Teaching Teachers for the Future, Australia
UCD	User-centered design
VANZ	Virtual Australia and New Zealand Initiative
VFT	Virtual field trip
VGI	Volunteered geographic information
WebGIS	Web-based geographic information system

List of Figures

List of Tables

About the Editors and Authors

About the Editors

Osvaldo Muñiz Solari is a professor in the Department of Geography at Texas State University. His major research interests are new technologies for global collaboration, online learning methods, and international flows. He is member of the Commission of Geographical Education of the International Geographical Union and member of the US national section of the Pan American Institute of Geography and History.

Ali Demirci is an associate professor in the Department of Geography at Fatih University in Istanbul in Turkey. His major research interests are geospatial technologies, curriculums, and teacher training in geography education. He is the coeditor of the book titled *International Perspectives on Teaching and Learning with GIS in Secondary Schools* published by Springer in 2012.

Joop van der Schee is professor for geography education at VU University in Amsterdam and at Utrecht University in the Netherlands. His major research
304 interests are geographic literacy, the use of digital maps in education, and geography teacher training. He is cochair of the Commission on Geographical Education of the International Geographical Union.

About the Editors

Marsha Alibrandi has taught geography and social studies at Cape Cod Technical High School, UMass Amherst, North Carolina State University, and Fairfield University. Currently retired, she has written and coedited two volumes, several chapters, and articles and worked with schools in several US states directing and/or implementing GIS in middle and high school settings.

Thomas R. Baker is an instructional designer and technologist (Ph.D. Curriculum and Instruction, 2002) serving as an education manager at Environmental Systems Research Institute (Esri, Inc.) where he leads STEM education activities, online community, and educational research initiatives related to GIS in education. Tom is also an adjunct researcher in the Center for STEM Learning at the University of Kansas.

Robert S. Bednarz is a professor of geography at Texas A&M University. He is North American commissioning editor for the *Journal of Geography in Higher Education*. In 2013 the NCGE recognized an article in the *Journal of Geography* he coauthored with Jongwon Lee, Components of Spatial Thinking: *Evidence from a Spatial Thinking Ability Test*, as the best article for geography program development.

Sarah Witham Bednarz is a professor of geography at Texas A&M University and associate dean for academic affairs for the College of Geosciences. She served on the committee that produced the report from the National Research Council Learning to Think Spatially (2006) and cochaired the Geography Education Research Committee of the Roadmap for 21st Century Geography Education Project in 2011–2013.

Tine Béneker is lecturer for applied sciences at Fontys University in Tilburg and lecturer for geography in education and communication at Utrecht University, the Netherlands. She is member of the board of the Royal Dutch Geography Society. Her research interest is in the field of world-mindedness of students, futures education, and geocapabilities.

Lara Bryant is an assistant professor of geography at Keene State College, New Hampshire. She earned her master's in applied geography and a Ph.D. in geographic education from Texas State University. She has a range of experience in geographic education, teaching geography in both rural and urban secondary public schools and offering geography classes geared toward preservice education majors in her current position.

Che-Ming (Jeremy) Chen is associate professor of geography at the National Taiwan Normal University. His research and teaching interests include geographic education, e-learning, and geospatial technologies. His current research focuses on e-learning for high school geography. In the last 5 years, he has held over 60 workshops helping secondary geography teachers to use GPS, GIS, and Google Earth in their classrooms.

Qiu Fen Jade Chen is a research assistant at the Office of Teacher Education at the National Institute of Education, Nanyang Technological University, Singapore. She has worked on geography research projects involving Singapore students and student teachers. She has published on research in students' environmental knowledge, attitudes, behavior, and fieldwork experiences of beginning teachers in Singapore.

Melody Crenshaw is an assistant professor of geography at Northwest Vista College in San Antonio. She earned her Ph.D. in geographic education from Texas State University. Her research interests include international social and environmental justice issues, geographic education in the US national parks, and community college geography student engagement.

Ali Demirci is an associate professor in the Department of Geography at Fatih University in Turkey. His major research interests are geospatial technologies, curriculums, and teacher training in geography education. He is the coeditor of the book titled *International Perspectives on Teaching and Learning with GIS in Secondary Schools* published by Springer in 2012.

Karl Donert is director of the European Centre of Excellence, digital-earth.eu, and adjunct faculty at the Centre for GeoInformatics at Paris Lödron University, Salzburg, Austria. He is President of EUROGEO and a UK National Teaching Fellow and a member of Academia Europea. He is Vice President of the Council of Europe NGO Commission on Democracy, Human Rights and Global Challenges and Chair of the Working Group on Climate Change and Human Rights.

Tim Favier is a researcher at the Utrecht University in the Netherlands. In his research, he focuses on how teachers can use geo-ICT in geography education and on its learning outcomes. Next to his research position at the Utrecht University, he works as a teacher trainer at the Fontys University of Applied Sciences in Tilburg, the Netherlands.

Donna Goldstein has worked in the GIS field for more than 25 years and has been the GIS coordinator with Palm Beach County School District, West Palm Beach, Florida, since 2001. As the GIS coordinator, she manages GIS applications, teaches GIS to K–12 teachers, and has been instrumental in integrating GIS technologies into existing K–12 curriculum.

Inga Gryl is a professor in the Department of Geography and the Department of Primary Social and Science Education at the University of Duisburg-Essen. Her research focuses on current geomedia in educational settings, emancipation and participation through media, and primary geography education. She is vice chair of the Association for Geographical Education in Germany.

Niem Tu Huynh is a research fellow at the Association of American Geographers (AAG). She served as the research coordinator and coeditor of "A road map for twenty first century geography education." She has worked closely with inner city high school teachers in D.C. and Maryland as part of the My Community Our Earth. She is concurrently a career advisor in the Career Planning Service at McGill University.

Jung Eun Hong is an assistant professor of geography in the Department of Geosciences at the University of West Georgia. She earned her B.A. at Ohio State University, master's at San Diego State University, and Ph.D. at the University of Colorado at Boulder. She has broad interests in GIS education including elementary to higher education level. Her current research is focusing on promoting teacher adoption of GIS technologies.

Thomas Jekel is a faculty in the Interfaculty Department of Geoinformatics at the University of Salzburg, Austria. He holds a M.Sc. in geography and communication science with emphasis on social geography and geoinformatics and a Ph.D. in geography education from the University of Salzburg.

Injeong Jo is an assistant professor in the Department of Geography at Texas State University. She earned her doctoral degree in geography education at Texas A&M University in 2011. Her research interests include geography and spatial thinking education, geography and social studies teacher education, geospatial technologies in education, and assessment development in geography.

Joseph Kerski holds three degrees in geography. For 22 years, he served as geographer and cartographer at three US Federal Agencies, including NOAA, the US Census Bureau, and the USGS. He has taught at primary and secondary schools, in community colleges, and in universities. Since 2006, he has served as education manager for Esri, focusing on geospatial technologies and education.

Sandra K. Metoyer is a research associate with the Education Research Center at Texas A&M University. She is pursuing her Ph.D. in geography from the College of Geosciences at Texas A&M University. Her areas of experience include spatial thinking and the educational use of geospatial technologies. She has conducted professional development for higher education faculty and in-service teachers.

Osvaldo Muñiz Solari is a professor in the Department of Geography at Texas State University. His major research interests are new technologies for global collaboration, online learning methods, and international flows. He is member of the Commission of Geographical Education of the International Geographical Union and member of the US national section of the Pan American Institute of Geography and History.

Reed Perkins is the Carolyn G. and Sam H. McMahon Professor of Environmental Science at Queens University of Charlotte, North Carolina. He received his B.A. in philosophy and M.S. in resource policy from the University of Michigan and a Ph. D. in forest hydrology from Oregon State University. He was named North Carolina Professor of the Year by the Carnegie Foundation for the Advancement of Teaching.

Uwe Schulze is a research assistant at Goethe University Frankfurt, Department of Human Geography, Working Group of Geography and Didactics. His research focuses on learning and teaching with geospatial technology in higher education and in teacher education and training as well as digital geomedia and spatial thinking, competence modeling, and curriculum development.

Bob Sharpe is an associate professor of geography and environmental studies at Wilfrid Laurier University in Waterloo, Ontario, where he teaches urban geography, cartography, and geomatics. His interests include human geography; geospatial technologies and geographic education research; applications of geospatial technology in teaching; and the behavior of individuals and small groups in urban space.

Geok Chin Ivy Tan is associate professor of the Humanities and Social Studies Education Academic Group and Sub Dean, Practicum at the National Institute of Education, Nanyang Technological University, Singapore. She has taught as a geography teacher and has been the head of the Humanities Department in secondary schools. She has also been a Gifted Educational Specialist in the Ministry of Education.

Henk Trimp is geographer, photographer, information scientist, psychologist, and actor by training. From 1985 onward he specialized in digital maps and GIS, and he has trained teachers on these topics since then. He now has a privately owned company that provides digital educational services and web development.

Joop van der Schee is professor for geography education at Faculty of Geosciences, Utrecht University, Utrecht, The Netherlands. His major research interests are geographic literacy, the use of digital maps in education, and geography teacher training. He is cochair of the Commission on Geographical Education of the International Geographical Union.

Yao-Hui Wang received his Ph.D. in geography from National Taiwan Normal University. He has 13 years of experience in teaching geography in the senior high school and promoting GIS education at secondary level. Wang is active in creating and teaching courses with geospatial technologies. His research interests are a reflection of his expertise. His focus is on geographic education, e-learning, and GIS.

Chapter 1
Geospatial Technology in Geography Education

Osvaldo Muñiz Solari, Ali Demirci, and Joop van der Schee

Abstract The book is presented as an important starting point for new research in Geography Education (GE) related to the use and application of geospatial technologies (GSTs). For this purpose, the selection of topics was based on central ideas to GE in its relationship with GSTs. The process of geospatial practices (GPs) as the way to learn about GST and with GST marks the central ideas. The sequential presentation of chapters starts with a theoretical approach, followed by the use of GST at different levels in formal education. Teacher education represents an important third topic that analyzes professional development when learning about GST and using GST. The fourth topic focuses on the effectiveness of GST as a tool for teaching and learning in order to assess geospatial thinking in different educational settings. The exploration of trends and recommendations for future development are the final topic of this book. In one sentence, the book touches upon the most important issues on the use of GST in education and includes the most up to date information and discussions related to GST. Although GE is not in the front line of the Information and Communication Technology (ICT) revolution it cannot stay behind as modern education uses technology to prepare better citizens. The importance of this book relies on this basic and fundamental fact.

Keywords Geography education • Geospatial technologies • Geospatial practices

O. Muñiz Solari (✉)
Department of Geography, Texas State University, San Marcos, TX 78666, USA
e-mail: o.muniz@txstate.edu

A. Demirci
Department of Geography, Fatih University, Istanbul, Turkey
e-mail: ademirci@fatih.edu.tr

J. van der Schee
Faculty of Geosciences, Utrecht University, Utrecht, The Netherlands

Faculty of Earth and Life Sciences, VU University Amsterdam, Amsterdam, The Netherlands
e-mail: j.a.vander.schee@vu.nl

O. Muñiz Solari et al. (eds.), *Geospatial Technologies and Geography Education in a Changing World*, Advances in Geographical and Environmental Sciences,
DOI 10.1007/978-4-431-55519-3_1

1.1 The Importance of This Book

Asked to mention the biggest challenge in their work many geography teachers come up with the digital revolution: computers, internet, e-learning, Google Earth and GIS entering the classroom. Most teachers see it as the biggest change in the recent past and as the biggest challenge for the near future. Sitting in a geography class and fly virtually to New York, Abu Dhabi or Ulan Bator to get an idea how life at different places on planet earth looks like using beamer, computers, tablets or mobile phones and software like Google Earth and Skype, that is geography in optima forma. Combining satellite images with digital maps, YouTube movies and TED talks really can be a good learning experience. Conditio sine qua non; however, is a good teacher who can help the learners to structure information, inquiry and evaluate.

Seen the importance of the digital revolution for Geography Education (GE) the Steering Committee of the IGU Commission on Geographical Education decided to make geospatial technologies (GSTs) for GE one its focus points. An important aim is to stimulate research and development in this field and to help to exchange research results and good practices internationally. In recent years many articles and books have been written about GSTs and GE. These publications can be found in journals within GE like *International Research in Geographical and Environmental Education, RIGEO, Journal of Geography, Geographie und Ihre Didaktik* and *Teaching Geography,* but also and in journals outside GE like *Computers & Education.* The Steering Committee of the IGU Commission on Geographical Education wants to help newcomers in the field of GSTs and GE to get an overview of the state of the art in this field. The result is this book. Contributions of almost 30 authors from 10 countries are brought together in 17 chapters. Not the last word as the world and its technology are changing fast, but as we hope an important intermediate point and a good starting point for new research and developments.

1.2 The Process of Selecting Topics to Discuss Central Ideas

GSTs and GE have a strong common ground in the twenty-first century education. Facing a tremendous impact and change with the advent of Internet and Web 2.0, GE has been forced to reconsider its education strategies within the digital networking arena. Information and communication technology (ICT) allowed us to study the earth in its physical dynamic and the world in its human and economic relationship. Data transformation, digital information and interpretation, time/space shrinking impact, and highly relational visual representations show some of the hidden challenges of GST to GE in the geo-enabling environment.

Central ideas to GE in its relationship with GST are the process of geospatial practices (GPs) as the way to learn about GST and with GST. We agreed on presenting first, certain level of analysis regarding the theoretical base which allows readers to understand what concepts and reasoning are important in GE as related to GST. Spatial thinking, geographical thinking, and geospatial thinking are part of this analysis in GE close related to the technological pedagogical content knowledge. However, to make a concept of spatial citizenship viable when dealing with GST in informal learning the concepts of space are also placed in discussion. Within the accepted absolute space the social space intervenes by creating permanent relational interpretations. The first part of this book introduces these clashing perspectives.

We all agree that the implementation of GST is not only important with new curricula in formal education but also through the complex and increasing practice in informal education. The second part of our book was set to respond to this implementation to see how GE in middle and high school as well as its applications in higher education is performed. Being aware of thousands of individuals, if not millions, engaged in crowd collaborative efforts and practicing with GSTs to resolve problems, place our work in a state of awareness. Informal education is worth the attention as a new trend of knowledge acquisition and we, the editors, consider this as an important task.

Even within the context of a constructivist perspective, teacher education for GST constitutes the first step to generate a sound learning process. The decision to consider teacher education as the third part of the book did not create any discussion. There is a firm agreement on the need for improvement when teacher professional development faces the difficult task of integrating pedagogical practice with technological development. The geo-enabling world paves the way to enhance teacher education yet it creates, with renew emphasis, special demands and efforts to learn new tools and techniques.

How effective is GP in education? This is a question to be answered by evaluation and assessment in our fourth part of the book. We estimate that this section has equal importance compared to the preceding parts. The need for geospatial thinking requires the effort of worldwide research to determine the level of its development. Furthermore, it is without saying that the application of valid and solid geospatial assessment practices has been sluggish. Training seems to be of vital importance when compared to technology availability.

GST has to be understood as a composite of tools and mechanisms to create new ways to teach and learn geography. Ultimately, it is about and with GSTs through GPs to improve the effectiveness of our interpretation and use of the earth. The educational process that is part of this symbiosis with technology has been delivered here as a set of ideas starting with some theoretical propositions; followed by various approaches on the implementation of GST. As a result, we believe that constant and increasing synergy is being built between GST and GE that carries important challenges which in turn oblige us to propose recommendations.

1.3 The Sequential Presentation of Chapters

The book, comprising 17 chapters in five parts, begins with a theoretical background, followed by chapters on implementation, on teacher education for GSTs, and on evaluation and assessment to end with a chapter on trends and recommendations for the future. Titled "Geospatial Practices. Theoretical Background", part one contains three chapters. Chapter 2 looks at the needs of GE in the twenty-first century and at how these needs can be met with GST by discussing opportunities and challenges that might be faced in the future. One of the most important contributions of GST to GE is its power to enhance spatial thinking skills. Chapter 3 explores the current position of spatial thinking in education including basic definitions, developments, and assessments in teaching and learning. The use of geoinformation (GI) in everyday settings has been increased. Chapter 4 explores the widespread use of GI in society using the concept of Spatial Citizenship and introduces a set of competences as well as a curriculum for in-service teachers to teach their students in secondary school classes how to become a spatially award society.

One of the important aspects of the book is that it explores the use of GST in formal and informal educational settings. Part two in the book evaluates the use of GST at different educational levels starting with middle and high schools and ending with higher and informal education. Chapter 5 explains why middle school is an extremely age-appropriate opportunity for students to gain many versatile skills through GST and describes theory, research and practice specific to the integration of GST into middle schools. Chapter 6 introduces different teaching methods and effective practices to integrate GST into high school geography curriculum. Chapter 7 discusses the roles of GST and the international trends in the application of GST in higher education exemplifying with a case study from a study abroad program in the islands of Micronesia. The proliferation of GST, many available from the Internet, especially in the last two decades has enhanced the opportunities to learn through informal ways. Chapter 8 discusses the importance of informal learning (IL) about GST and with GST, using a global example represented by volunteered geographic information (VGI).

Professional development for teachers has always been among the most important concerns when it comes to effective use of GST in teaching and learning. Titled "Teachers Education for GST" part three in the book approaches this issue in three chapters. Chapter 9 explores the growing educative role of WebGIS conceptually, technically, and practically and explains how learners and educators can utilize WebGIS in their lessons by taking into consideration new choices and constraints provided mainly by GIS in a cloud-based platform. Chapter 10 evaluates major barriers prior to implementation of GST in classrooms with a specific focus on teachers and introduces six models for in-service teacher training programs for GST with a discussion on effective and useful in-service teacher training programs for the future. One of the most cited benefits for learning of GST is its power to support inquiry based learning (IBL). Chapter 11 explains the role of IBL in GE, and explores didactic models for integrating GIS in inquiry projects along with frameworks and a model for teachers to design and conduct such projects and to train the necessary teacher competencies.

It is important to evaluate the effectiveness of GST as a tool for teaching and learning in order to develop methods to assess geospatial thinking in different educational settings. The three chapters in part four assess the effectiveness of using GST in education. Chapter 12 first evaluates the potentials of GST for teaching and learning and then discusses whether geospatial practices are actually effective in education addressing different concerns and questions raised in the literature. Chapter 13 assesses the use of GIS in teaching in ten countries with the conditions encouraging and discouraging the integration of GIS usage based on 15 research articles published basically in the last decade. Chapter 14 reviews how geospatial thinking is assessed in high schools first by looking at the literature to identify recent research trends, and then by conducting an online survey on educators involved in high school GE from around the world.

The final part of the book explores the trends in the development of GST and makes some recommendations for the future in order to obtain more benefits and more effective procedures in the process of applying GST in education. Chapter 15 first discusses the opportunities by evaluating three converging trends, namely an awareness of global challenges, the spread of GST to the general public, and the geo-enabling of everything and then evaluates the technological, pedagogical, and administrative challenges for using GIS in teaching and learning. Chapter 16 introduces the Digital Earth concept and examines the changing place of education in it by identifying key stakeholders involved in the development of geospatial industry and their educational needs along with some suggestions of strategies that they should implement to achieve Digital Earth education. Chapter 17, the final chapter of the book, introduces a geospatial science agenda for GE for the future based on the existing and newly emerging opportunities and challenges for using GST in teaching and learning.

1.4 Why GST Matters for Education

GST refers to equipment used in visualization, measurement, and analysis of earth's features, including systems as Global Positioning Systems (GPS), Geographical Information Systems (GIS), Remote Sensing (RS) and digital globes. Many authors see great advantages of using GST in education as geodata availability is much quicker and easier and geodata processing is many times faster and accurate than it was before. Also the opportunities for visualization are huge. In a few decades computers and internet really have changed the world. Although GE is not in the front line of this revolution it cannot stay behind as modern education should prepare students for the world of today and tomorrow. For GE it offers many opportunities. GSTs have the potential to enhance students' twenty-first century skills and can stimulate a new way of learning or at least offer better opportunities to develop higher order thinking skills. Modern GST can help learners to gain a better view of the world and offers opportunities for learners to better understand planet earth and reflect on the near future.

The question is; however, does education keep in pace with the geospatial revolution? Most of the geospatial revolution takes place outside the classroom, not inside. Many young students are using GST while playing games and using all kind of applications on tablets and mobile phones. To get education more involved in this active process the role of the teachers is vital. Do they use GSTs in school and if so, how do they use them? Do they use it in the way they are used to when working with paper maps and schoolbooks? Or do they introduce new ways of learning like virtual fieldwork and inquiry projects using web atlases and GIS? Do they know how to maximize the advantages of GSTs not only in a technical way but also for higher learning goals? Finally, how can teachers use GSTs effectively to help students to reflect about phenomena and processes on planet earth?

GSTs have a big potential but they are tools, not more and not less. Modern geographers in education that ignore GSTs so far should have a try and eat a piece of the cake. Not only by reading chapters in this book but, even better, by participating in projects in which students use GSTs like GPS and GIS.

1.5 How This Book Will Be Helpful to Use GST for Education

The target group of the book is broad. It addresses teachers, professionals, scholars, and policymakers who are interested in using GST in education. The book will help teachers in primary and secondary schools as well as professionals in higher education to learn different strategies, methods, and approaches to incorporate GST into their work environment. It provides, at the same time, with some examples and case studies from different countries to understand real problems. The book touches upon the most important issues on the use of GST in education and includes the most up to date information and discussions.

Understanding the recent developments and discussions in this field, especially in the academic world is crucial for scholars to conduct research effectively. The book, therefore, is useful for scholars who are interested in teaching and learning with GST to broaden their understanding, follow the most recent research results and discussions, and become aware of the areas where further or new research is needed. Another important target group of the book is the policymakers in the education business. The book will be helpful for the policymakers to understand that GST is developing very fast and provides immense opportunities to educate society. Reading the book, policymakers will understand that GSTs are already present in our everyday lives. All policymakers need to do is look around. They will understand the power of these technologies and will want to utilize them for education. Including many examples from around the world, the book will also provide policymakers with a range of effective strategies they can use to benefit GST in education.

1.6 Challenges and Recommendations

The decision to close this book with an analysis of trends and recommendations was based upon the belief that opportunities and challenges are equally necessary to measure as a result of a geo-enabled world with a constant transformational technology in place. As part of this revolution, GSTs impact and transform formal education and informal learning environments alike, yet geo-capacity building in education is still behind unless trans-disciplinary strategies to learning and teaching are accomplished. In this regard, we were right in our decision to pursue a set of recommendations. There are no strategies in place to initiate new curricula in secondary and higher education for increasing geospatial capacity building; tactics that require an organized as well as consistent thrust to guide teachers and students about how to learn GSTs and use them to conduct efficient GPs.

We agree with our contributors to this book that the world faces complex challenges which are global in nature yet creating constant and permanent impacts on our daily individual lives. These challenges are intermixed with increasing technological development that forces individuals, communities, and society as a whole to acquire new knowledge overloaded by geospatial components. What is interpreted as the landscape of GSTs in education characterized by a full spectrum of opportunities is also recognized as plenty of challenges that instigate recommendations of equal importance.

Recommendations vary among important issues, some of which demonstrate the heavy weight that GST practitioners are carrying in formal education. Curriculum is one these issues that places great emphasis on the process of spatial thinking as the major and sole preoccupation. Another issue is the need to resolve bottlenecks concerning abilities to master GST integrative skills by teachers or facilitators participating in active and collaborative learning. It is not only geospatial literacy an important issue to tackle as crucial to build teacher capacity but also the reduction of the digital divide between new and older generations. For this purpose the concept of community of practice plays an undeniable role to create collaborative environment.

Part I
Geospatial Practices. Theoretical Background

Chapter 2
Digital Geography Education in the Twenty-First Century: Needs and Opportunities

Joop van der Schee, Henk Trimp, Tine Béneker, and Tim Favier

Abstract The introduction of geospatial technologies is changing geography education very fast. Google Earth, web atlases and many location based services are available from the Internet and offer the opportunity to study almost every place in the world anytime anywhere. This opens up a great perspective for geography education in a way not known before. Nevertheless also in geography life is not just a bed of roses. The introduction of geospatial technologies is not always simple, due to technical problems and fast changing devices, a lack of experience in teaching with geospatial technologies and a huge amount of information that may prevent learners to see the wood for the trees. However, the advantages of using geospatial technologies in geography education far outweigh the problems if geographical thinking is involved. This contribution discusses briefly the needs and opportunities of digital geography education in the twenty-first century.

Keywords Geography education • Geospatial technologies • Geospatial thinking

2.1 Introduction

This chapter will start with a sketch of the core business of geography education. Then we will have a short look at the recent history of geospatial technologies in geography education to show where we come from and how fast the world of

J. van der Schee (✉)
Faculty of Geosciences, Utrecht University, Utrecht, The Netherlands

Faculty of Earth and Life Sciences, VU University Amsterdam, Amsterdam, The Netherlands
e-mail: j.a.vander.schee@vu.nl

H. Trimp
Anaximander, Amsterdam, The Netherlands
e-mail: henk@anaximander.nl

T. Béneker • T. Favier
Faculty of Geosciences, Utrecht University, Utrecht, The Netherlands

Geography Department, Fontys University of Applied Sciences, Tilburg, The Netherlands
e-mail: t.beneker@uu.nl; t.favier@fontys.nl

O. Muñiz Solari et al. (eds.), *Geospatial Technologies and Geography Education in a Changing World*, Advances in Geographical and Environmental Sciences,
DOI 10.1007/978-4-431-55519-3_2

teaching with geospatial technologies is changing. Next we will discuss some opportunities and challenges we might face in geography education using geospatial technologies. In the following chapters these opportunities and challenges will be explored further.

2.2 Geography Education

One of our dreams as geography educators is that all over the world young people investigate and evaluate their living environments to discuss the challenges and future perspectives of these environments. And that they do this supported by their parents and their teachers, using modern means of communication and geospatial information technology. Sometimes we see signals that this dream comes true. School children from all over the world participate in environmental protection programs like DeforestACTION. This global learning project enables children to monitor Borneo's rainforest from drones and satellite images helping to protect it from illegal loggers (Fig. 2.1). Rapid illegal rainforest logging is escalating driven by an insatiable demand for palm oil from China, India and elsewhere. Palm oil is the world's most consumed vegetable oil. School children are becoming involved to help save endangered animals and their home, the rainforest. This hands-on approach of ecology and conservation-in-action empower young world citizens. DeforestACTION shows that together they can have a voice and play an active role in protecting the environment and our planet. Geography education will help students to raise relevant questions as where is deforestation taking place, at what scale and speed (where?), how does it happen and who is doing it (what?), why does it happen (why?), who is benefitting and at the expense of whom and what (consequences?), what are the alternatives (policy?), and how does this all relate to myself and our society (awareness and commitment?). Geography education is about meaningful learning starting from geographic questions. It is the fascinating story of people that live on planet earth at different spots in different ways in conditions that change continuously. Morgan (2013: 275) analysing a text of Mackinder states that

> To think geographically is to have a trained capacity to construct a mental map to see patterns, to recognise relationships, to see movement, to take that map and 'clothe it in meaning.

The digital revolution offers more and more fascinating possibilities to discover planet earth. Computer technology, the Internet and Global Positioning Systems (GPS) underlie recent innovations in the field of geospatial technologies (GSTs). Geospatial technology (GST) refers to equipment used in visualization, measurement, and analysis of earth's features, typically involving such systems as GPS, Geographical Information Systems (GIS), Remote Sensing (RS) and digital globes

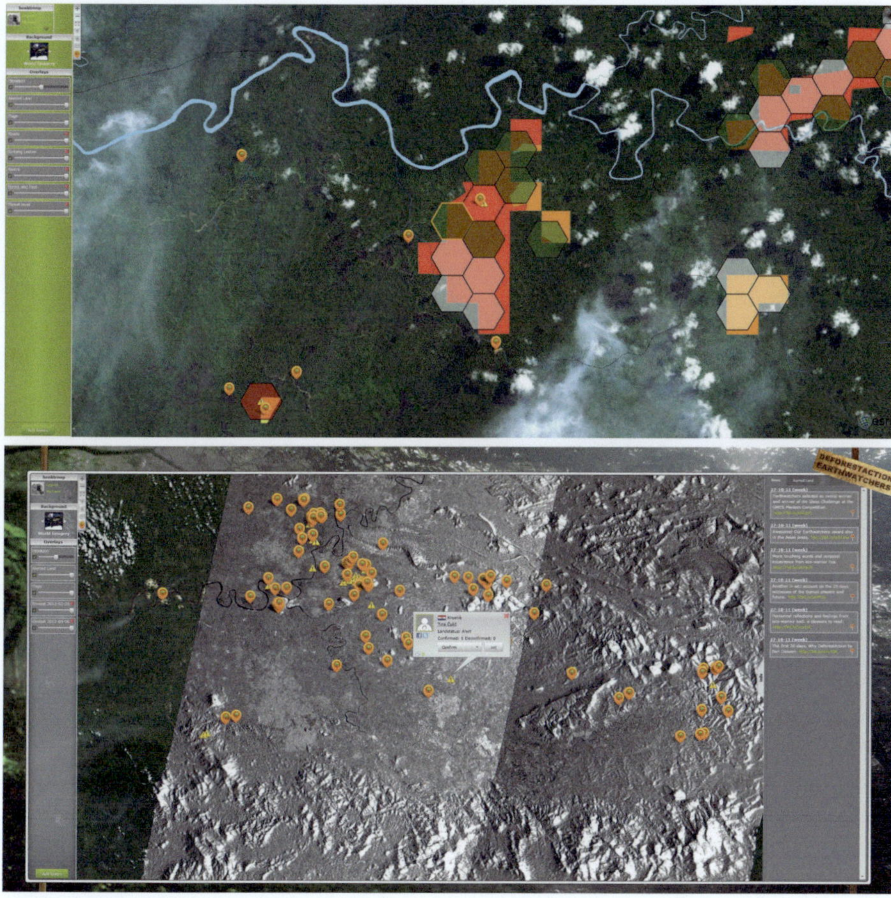

Fig. 2.1 Map of Borneo used by DeforestACTION (Screenshots van de Earthwatchers app; http://earthwatchers.cloudapp.net/)

(Cimons 2011; Baker et al. 2014). Geospatial thinking and GSTs can be situated in a framework of pedagogy, ICT, geography and science (Fig. 2.2).

Spatial thinking can be defined as a set of abilities to visualize and interpret spatial concepts like position, distance, relationships and change through space. Geospatial thinking is a specialized form of spatial thinking, focusing on patterns and processes that take place on or near the earth's surface (Baker et al. 2014). Geospatial thinking is also a specialized form of geographical thinking. GSTs are more and more the tools to help us in the process of thinking and learning about what is happening on planet earth.

But how useful and often fascinating the GSTs are, they are just tools. The core business for geographers is the questions they ask about what is happening on planet earth. Geography is concerned with human- environment interactions in the context of specific places and locations (Haubrich 1992), and with issues that have a

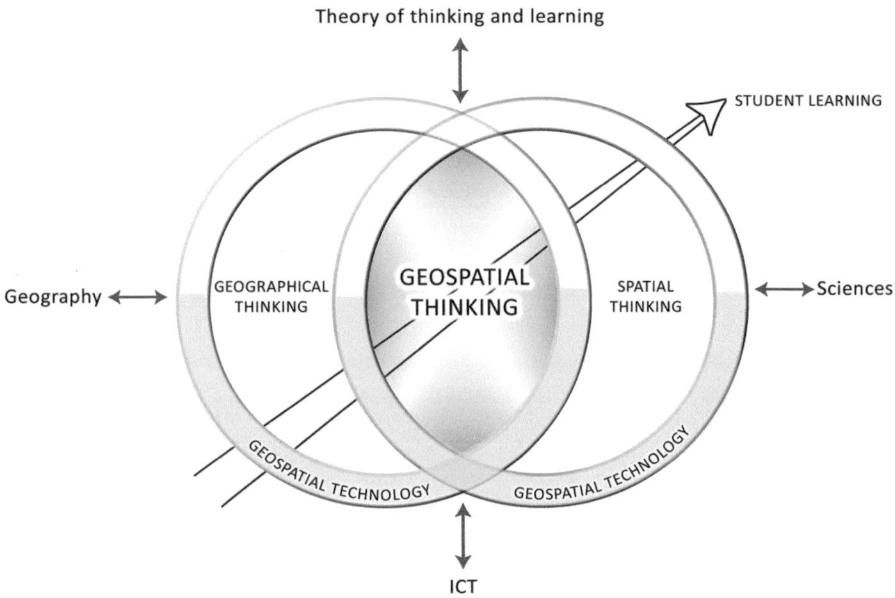

Fig. 2.2 A model to position geospatial thinking and geospatial technologies

strong geographical dimension like natural hazards, climate change, energy supplies, land use, migration, urbanisation, poverty and identity. Geography is a way of seeing places and thinking geographically to make sense of the world (Butt 2011). Using core or key concepts systematically helps us to think geographically (Geographical Association 2009). Place and space are the overall core concepts in geography. Every location and region is a place with its diversity in identity, history, people, and nature.

Every place (the 'local') is related to other places and regions, and moreover part of bigger regions (or countries) and the 'global'. Of course places are social constructions of people and can be seen from different perspectives. Looking from a spatial perspective the focus is on flows, mobility, relations, interactions, spatial pattern and networks, for example in trade, migration and communication. Place and space are supported by other key concepts, such as diversity, interdependence, scale, change and interaction (Taylor 2008; Lambert and Morgan 2010).

Geography education focuses on human-environment interactions and how they work out in different ways at different locations and helps us to get an overview of how life on planet earth goes on. GSTs offer 'Borneo-watchers', woodcutters, business men and policy makers unique opportunities to see what is going on and show that without geography we cannot make sense of the modern world nor make plans for its future.

2.3 Old and New in GST

GST started a very long time ago when people tried to explain a location or a good travel path to other people using a stick sketching a map in the earth or using stones to point out positions and routes. If you ask someone nowadays to show you the way, e.g. in Tokyo the shortest route to Asakusa from Kototoi Dori, big chance he or she will use a mobile phone to point out the direction on a digital map. Although some basic principles in learning with maps are still the same as a long time ago, today our instruments and our knowledge base have changed enormously. The use of GST is well-known and widespread in the military and hazard management, but its influence is pervasive everywhere, in such areas as land use, e-farming, retail planning and environmental protection.

From stick and stones via handmade maps and printed maps to digital maps was a long way. Computer hardware and software enabled the big leap forward from printed maps to digital maps. The first mechanical computers were developed in the nineteenth century, the first personal computers were introduced in 1980 by Sinclair Instrument Ltd. and IBM in 1981 (Fig. 2.3) and the Internet started to become available for the general public after 1992, when the introduction of the graphical web browser Mosaic meant a breakthrough of the hypertext-based World Wide Web.

After the invention of the printing press the introduction of the computer in combination with the Internet is a second big revolution in the world of learning and communication. "On our screens, on our phones, in our textbooks and magazines, our images of the world are changing faster than the world is itself" (Dorling 2012).

The terminology in the world of geo-information changes almost as fast as the development of the technical innovations itself. Many different terms alternate like New Media, Geo-ICT, GIS, Geographic Information Science, geomedia, geodesign, Volunteered Geographic Information (VGI), Web 2.0, Public Participatory GIS (PPGIS) and Neogeography. Until recently GIS was the most used word in geographical education for the activities connected with describing, analyzing and presenting digital geodata. GIS is defined by Burrough (1986) as "a collection of tools for collecting, storing, and visualizing spatial data about the world around us". More recently we see more and more user-generated content. The Internet is used by individuals to create and present their own world. This is where the term VGI comes in. Goodchild (2007) states that VGI has the "potential to be a significant source of geographers' understanding of the surface of the Earth". Citizens as sensors of what is happening on planet earth. A group of individuals collecting data responding to needs of a community or sometimes just for a hobby or fun. Although Goodchild sees a lot of advantages of the use of GSTs, for instance in hazard management, he also warns us to be critical. The reliability of data, the privacy of data, the digital divide and the vulnerability of the internet are important issues.

Borruso (2013) discerns nowadays a second revolution after the GIS revolution started during the last decades of the twentieth century. During the first revolution

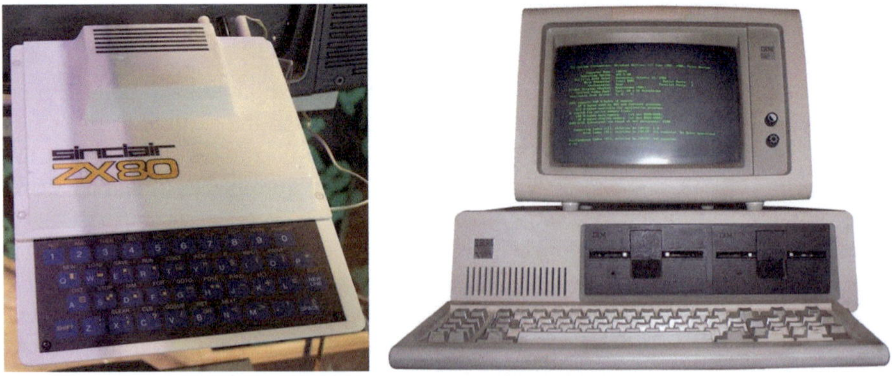

Fig. 2.3 Sinclair ZX80 (1980) and first generation IBM PC (1981)

paper maps were replaced by digital maps that were used in inquiry-driven settings to foster knowledge acquisition. However, to analyze change temporally and spatially dynamic maps are required. Decision-makers use dynamic web maps to solve problems on a daily basis (Kerski 2013: 12/13). Digital maps evolved in cloud-based shareable maps. According to Borruso the so called Neogeography starts at the beginning of twenty-first century with citizens creating their own geography. This second revolution is made possible by the availability of a huge amount of free data, the fast diffusion of advanced laptops, tablets and mobile phones, low cost high speed internet with more bandwidth and a growing group of skilled computer users. In addition to VGI and Neogeography the name Web 2.0 is used to indicate that Internet users not just download information from the Internet but also interact and collaborate with each other using social media like blogs, wikis and Facebook. So what we see is hardware innovations, a fast growing set of easy available data and recently more and more communities of users that also create and present their own data.

2.4 New Opportunities and Challenges for Geography Education

For many of us using a mobile phone with GPS and geo-browsers such as Google Earth is as normal as eating every day. Life outside school without GSTs is for many of us inconceivable. Inside schools we see big differences in the use of modern GSTs.

Computer software offers great opportunities for learning and teaching geography as students and teachers have a lot of geographical data at hand in satellite images, maps and videos. As GSTs can be used in an interactive way they can also be very helpful in an educational system that advocates help students with different abilities. The possibilities to find and to handle information in an interactive and very fast way are huge and booming. However, not everybody is optimistic about the new software that is available for geography education. Critical voices about the use of GSTs in education focus on the different weaknesses. First, in many schools in the world in less developed countries there are no facilities for using computers in the classroom. Hardware, software or even electricity is missing. Secondly, in more developed countries GSTs are entering classrooms slowly and often only recently as spatial technologies are not or are not well integrated in the curriculum and exams. Thirdly, most GIS and RS packages are not made to be used in education and many new geospatial technologies change so fast that it is not easy for teachers to find the time and energy to keep up with the latest and best innovation in a school context. And last but not least, most teachers are not trained to use GSTs in schools. They lack not only technological knowledge but also technological pedagogical content knowledge (TPCK) (Mishra and Koehler 2006). Of course the reality of GSTs in education is black nor white but everywhere different and changing. The use of smartphones is growing fast all over the world, web based GIS is less complex than it was, and a new generation of teachers that are digital natives is coming in. The more we are flooded by data and modern devices, the more urgent is the need to raise questions about the use and reliability of data. More important even is to find good ways to help students to give meaning to the fast growing amount of information in an era of fast developing technology. This brings us to the following needs and opportunities for training and research.

1. Practicing higher order thinking.

Outside schools young people are familiar with using geo-technologies but not in higher order thinking skills like structuring and analyzing (digital) information. This could be the added value of spatial thinking or geospatial thinking. Jo and Bednarz (2009) developed a tool to rethink the use of GSTs in geography education. Their taxonomy for spatial thinking has three dimensions: (1) concepts of space, (2) processes of reasoning, and (3) tools of representation. The taxonomy was developed and used to evaluate questions in four US high school level geography textbooks. The results indicate that textbook questions focus on low-level spatial concepts and textbook questions only rarely encourage higher-order cognitive skills. The study makes clear that it is good to ask the question what it exactly means to foster learning by thinking spatially or geospatially. Using GSTs is no guarantee for learning higher order thinking skills. The development of conceptual geographic frameworks to organize and structure information seems to be more relevant than ever. The question is how we can use GSTs to stimulate higher order thinking skills.

2. Meaningful learning using GSTs for studying regional systems.

Uhlenwinkel (2013) states that the competence of thinking geographically needs more attention in the field of GIS and spatial thinking. Information processing skills are often central in spatial thinking. If we focus more on thinking geographically ‘answering a geographical question’ and ‘giving meaning to the information’ are added explicitly to information processing. This can help to simulate higher order thinking skills as students learn to study not only facts and concepts but also geographical relations within and between regions and regional systems. Kerski (2013: 25/26) states that asking questions and being inquisitive are critical to the successful use of web maps and GIS in education: "Through the use of these web mapping technologies, instructors can help students to begin analyzing the "whys of where" – the essence of geographic inquiry". Kerski emphasizes that asking about the whys of where is not the end of the story: "Students need to ask and grapple value based questions"……"This captures not only the heart of the spatial thinking, inquiry, and problem-based learning, but of education for activism – to make a difference in this changing world of ours".

The advantage of using GSTs to study regional systems is that they offer huge amounts of up-to-date data in a flash, often even in an interactive way. Google Earth, Street View, YouTube, Panoramio and other tools make it possible to zoom in on areas all over the world on a computer screen in 2D or 3D. In addition, web atlases can help to see patterns and processes in regions. GSTs do more than just give information. They also include tools to analyze information, e.g. to select certain phenomena or areas, to investigate relations between different phenomena within and between regions and to investigate and evaluate effects of decisions in an interactive way. Moreover, GSTs offer tools for the learner to be active and creative and GSTs are very helpful in visualizing plans and scenarios. So using GSTs we have at our disposal a well-visualized, easy, fast and interactive way to analyze and evaluate regional systems and regional development. Doing so, learners need to use higher order thinking skills.

The use of GSTs in education can be illustrated in different ways. As recordings of hurricanes and other disasters can be very shocking and help students to understand the impact of the power of nature, geospatial information available via GIS can help students to understand the causes of the event and to frame geographical learning (Sinton and Bednarz 2007; Fargher 2013). Other examples of the successful use of GSTs come from the local domain. Borián (2012) describes a project in which students in different European countries investigate, compare and explain the water quality in their local area. Using BISEL (Biotec Index at Secondary Education Level) students measured water quality and exchanged the results in a set of digital map layers. In conferences and through the internet students discussed the different results and geographical conditions. Fargher (2013) writes that PPGIS is "a promising way to further develop geographical understanding in schools of local issues because it attaches importance to deep local knowledge being of value in society".

Although many researchers, teachers and students report very positive about the use of GSTs there are some questions. First of all, there is not much evidence from research about the positive effects of using GSTs yet. Systematic studies in this field are necessary. Secondly, what we know so far is that students need more help and structure to frame the information they encounter. Favier and Van der Schee (2014) developed geography lessons about water management with GSTs using a serious game and a web GIS and compared them with conventional geography lessons that had the same content. The research data showed that the lessons with GSTs were motivating for the students and contributed significantly more to the development of students' geospatial relational thinking than the conventional lesson series. However, higher order thinking skills were only partly acquired as most students had difficulty to evaluate the system of water management and its variables after the set of lessons. More structure provided by the teachers seems to be indispensable.

3. Teachers and their technological pedagogical content knowledge (TPCK).

Good geography teachers are needed to help students obtain geospatial thinking skills using GSTs. For innovations in education the teacher is the crucial factor and whether he or she will apply GSTs in teaching and learning will depend on external factors like the perceived need to do so and the perceived manageability as well as on internal factors like his or her professional geographical knowledge and motivation (Bednarz and Van der Schee 2006). Teacher training courses that focus on TPCK are still scarce, while we need teachers that can help students to see the wood for the trees and acquire the skills to ask critical questions about a fast changing world. The question is not whether to use GSTs or not but how to use them in a thoughtful and critical way.

2.5 Conclusion

GST offer rich opportunities for geography education and related disciplines. The danger using GST is to stick at a low level, the level of interesting pictures and maps. The challenge is to go beyond. Developing meaningful learning units that help to train students' higher order thinking skills is what we need to make optimal use of GST. Teachers play an important role here. That their geographical knowledge and understanding is crucial cannot be stressed enough.

References

Baker, T., Battersby, S., Bednarz, S., Bodzin, A. M., Kolvoord, B., Moore, S., Sinton, D., & Uttal, D. (2014). A research agenda for geospatial technologies and learning. *Journal of Geography* (online), *1–13*.

Bednarz, S. W., & Van der Schee, J. A. (2006). Europe and the United States: The implementation of geographical information systems in secondary education in two contexts. *Technology, Pedagogy and Education, 15*(2), 191–206.

Borian, G. (2012). GIS in natural science teacher training. In A. Milson, A. Demirci, & J. Kerski (Eds.), *International perspectives on teaching and learning with GIS in secondary schools* (pp. 125–130). Dordrecht: Springer.

Borruso, G. (2013). Web 2.0 and neogeography. Opportunities for teaching geography. *Journal of Research and Didactics in Geography (J-Reading), 2*(2), 43–55.

Burrough, P. A. (1986). *Principles of geographic information systems for land resources assessment*. Oxford: Clarendon.

Butt, G. (2011). Introduction. In G. Butt (Ed.), *Geography, education and the future* (pp. 1–11). London: Continuum.
Cimons, M. (2011). *Geospatial technology as a core tool*. US News provided by National Science Foundation. Downloaded June 25, 2014, from www.usnews.com/science/articles/2011/05/11/geospatial-technology-as-a-core-tool
Dorling, D. (2012). Mapping change and changing mapping. *Teaching Geography, 37*(2), 94–98.
Fargher, M. (2013). Geographic Information (GI) – How could it be used? In D. Lambert & M. Jones (Eds.), *Debates in geography education* (pp. 206–218). London: Routledge.
Favier, T., & Van der Schee, J. A. (2014). The effects of geography lessons with geospatial technologies on the development of high school students' relational thinking. *Computers & Education, 76*, 225–236.
Geographical Association. (2009). *Geography: A different view*. Sheffield: Geographical Association.
Goodchild, M. F. (2007). Citizens as sensors: The world of volunteered geography. *GeoJournal, 69*(4), 211–221.
Haubrich, H. (1992). *International Charter on Geographical Education*. Freiburg: Commission on Geographical Education, IGU (see also www.igu-cge.org).
Jo, I., & Bednarz, S. W. (2009). Evaluating geography textbook questions from a spatial perspective: Using concepts of space, tools of representation, and cognitive processes to evaluate spatiality. *Journal of Geography, 108*(1), 4–13.
Kerski, J. J. (2013). Understanding our changing world through web-mapping based investigations. *Journal of Research and Didactics in Geography (J-Reading), 2*(2), 11–26.
Lambert, D., & Morgan, J. (2010). *Teaching geography 11–18*. Maidenhead: Open University Press.
Mishra, P., & Koehler, M. J. (2006). Technological pedagogical content knowledge: A new framework for teacher knowledge. *Teachers College Record, 108*, 1017–1054.
Morgan, J. (2013). What do we mean by thinking geographically? In D. Lambert & M. Jones (Eds.), *Debates in geography education* (pp. 273–281). London: Routledge.
Sinton, D. S., & Bednarz, S.W. (2007). About that G in GIS. In D. S. Sinton & J. Lund (Eds.), *Understanding place GIS and mapping across the curriculum*. Redlands: ESRI Press.
Taylor, L. (2008). Key concepts and medium term planning. *Teaching Geography, 33*(2), 50–54.
Uhlenwinkel, A. (2013). Spatial thinking or thinking geographically? On the importance of avoiding maps without meaning. In T. Jekel et al. (Eds.), *GI_Forum 2013. Creating the GISociety* (pp. 294–305). Berlin: Wichmann.

Chapter 3
Spatial Thinking in Education: Concepts, Development, and Assessment

Sandra K. Metoyer, Sarah Witham Bednarz, and Robert S. Bednarz

Abstract Spatial thinking has always been a fundamental cognitive skill for competency in geography. However, interest in it has increased in recent years as technological advances have driven political and societal changes producing a renewed awareness of its importance. This is especially true in the context of geospatial technologies (GST). The growth, expansion, and power of GST demands a citizenry with well-developed spatial thinking skills. But research exploring spatial thinking in an educational context is scant.

This chapter explores the current position of spatial thinking in education. First, we describe existing research in spatial thinking. We focus on advances in technology which have led, in part, to the increased interest in the topic. The roles of spatial thinking and GST in curricula are explored. Promising methods for assessing students' spatial thinking are reviewed in order to provide guidance for curriculum decision-making. The chapter concludes with a summary of the current state of spatial thinking in education and with recommendations for further research.

Keywords Spatial thinking • Geography education • Geospatial technologies

3.1 Spatial Thinking in Education: Concepts, Development, and Trends

Spatial thinking has always been a fundamental cognitive skill in geography. Space is a key organizing concept for our discipline. Moreover, geographers use spatial thinking supported by spatial representations such as maps to: pose geographic questions; collect, organize, and analyze geographic information; and explain and

S.K. Metoyer (✉)
Galveston College, Galveston, TX, USA
e-mail: smetoyer@gc.edu

S.W. Bednarz • R.S. Bednarz
Texas A&M University, College Station, TX, USA
e-mail: s-bednarz@tamu.edu; r-bednarz@tamu.edu

O. Muñiz Solari et al. (eds.), *Geospatial Technologies and Geography Education in a Changing World*, Advances in Geographical and Environmental Sciences,
DOI 10.1007/978-4-431-55519-3_3

communicate geographic patterns and processes—practices critical to the development of twenty-first century competencies (Bednarz et al. 2013). Technological factors including the explosion of location-based geospatial technologies (GST), such as geographic information systems (GIS), and political and social forces have combined to drive a renewed awareness of, and interest in, spatial thinking. These developments have also contributed to "the spatial turn" in geography (Goodchild and Janelle 2010).

This chapter explores the current position of spatial thinking in education. First, we briefly review definitions of spatial thinking and four conceptualizations of how spatial thinking develops in individuals. Second we describe the political, social, and technological factors driving interest and research in spatial thinking in education. In the third section we examine efforts to assess spatial thinking that may provide evidence to guide curriculum decisions regarding the teaching and learning of spatial thinking. The chapter concludes with recommendations for further research.

3.2 Definition of Spatial Thinking

Spatial thinking can be defined as a constructive combination of cognitive skills comprised of knowing concepts of space, using tools of representation, and applying processes of reasoning (NRC 2006, p. 12). Spatial thinking allows people to use space to model the world (real and theoretical), structure problems, find answers, and express and communicate solutions. The inclusion of concepts of space makes spatial thinking unique from other types of thinking (NRC 2006). Concepts of space are declarative forms of knowledge, the building blocks for spatial thinking. Location, dimensionality, continuity, pattern, spatial association, network, and proximity are examples of spatial concepts that have been explicitly recognized by researchers (Gersmehl and Gersmehl 2007; Golledge 2002; Janelle and Goodchild 2009). Tools of representation such as maps, graphs, sketches, diagrams, images, and models enable and support spatial thinking. They are used in a variety of modes (mental images, visual media, tactile, auditory, and kinesthetic forms) to identify, describe, explain, and communicate information about objects and their associated spatial characteristics (NRC 2006).

Spatial thinking often necessitates complex reasoning (Jo and Bednarz 2009). Reasoning is the capacity of individuals to think, make sense of the world, and understand. Processes of reasoning are crucial for learning as individuals obtain, change, or justify practices, institutions and beliefs (Kompridis 2010). Processes of reasoning include low levels of thinking, such as recognizing, defining, and listing, and higher levels of thinking, such as evaluating, synthesizing, and generalizing (Jo and Bednarz 2009).

3.3 Conceptualizations of Spatial Thinking

Understanding spatial thinking in terms of human development and learning is a necessary precursor to discussions of current trends, and interest, in teaching spatial thinking. One of the most important debates informing spatial thinking in education concerns the extent to which spatial thinking is, in some sense, innate. Many researchers have attempted to conceptualize the development of spatial thinking (Allen 2003; Kuipers 1978; MacEachren 1992; Montello 1998; Newcombe and Learmonth 2005). These theories can be grouped into four broad categories: nativist, Piagetian, Vygotskyan, and interactionist (Kim et al. 2012). Nativists argue that children are born with a biologically determined level of spatial thinking, and, even though spatial thinking may develop with age and experience, biology pre-determines ability. Contrary to the nativist approach Piaget argued that "infants are born without knowledge of space, and without a conception of permanent objects which occupy and structure that space" (Newcombe and Huttenlocher 2003). Piagetians propose a sequential progression of understanding from topological space to projective and Euclidean space (Piaget and Inhelder 1948). The egocentric-to-allocentric shift predicted by the Piaget approach has inspired a large volume of research related to spatial thinking in education (Downs and Liben 2001; Golledge et al. 1993; Shelton and McNamara 2004; Thommen et al. 2010). Both the nativist and the Piagetian conceptualizations of spatial thinking minimize or ignore the social and cultural influences on humans' development of spatial thinking, for example, the role played by cultural tools such as language or maps. Those who view spatial thinking with a Vygotskyan (or sociocultural) conceptualization emphasize these social and cultural influences on individual intellectual development and encourage the examination of cultural tools and environments that affect human development (Gauvain 2008). The interactionist conceptualization of spatial thinking asserts that components of each of the previously discussed approaches are valid; newborn children likely arrive with a set of biologically determined innate spatial abilities as nativists argue, children and novices show predictable developmental transitions as a Piagetian approach would argue, and the influence of life experienced through culture and cultural tools is clearly evidenced in the variance of spatial thinking observed across individuals and cultures (Newcombe 2000). The interactionist approach argues for the influence of both nature and nurture on the cognitive development of spatial thinking.

Educational approaches for fostering spatial thinking typically utilize an interactionist approach because it recognizes that individuals have different starting points for spatial thinking, but spatial skills can be improved through training and scaffolding. An interactionist approach provides teachers and policy makers with the opportunity to consider a wide range of educational strategies. Even though students bring different spatial thinking approaches and preferences to the classroom, tools of representation paired with quality instruction can enhance and develop multiple strategies for spatial thinking.

3.4 Interest in Teaching Spatial Thinking

Geospatial technologies (GST), defined as technologies that facilitate visualization, measurement, mapping, wayfinding, or spatial analysis of features both concrete and conceptual on Earth's surface and subsurface, have become ubiquitous, and now location matters more than ever. Paper maps have been replaced by smartphones with digital maps, navigation systems, and global positioning units (GPS). Most vehicles are also equipped with these technologies. We report our location frequently through social media. Web-based mapping and analysis software give even a novice user access to a wide variety maps and to constantly growing functionality to display and analyze spatial data (see for example GIS Cloud at http://www.giscloud.com/). The widespread availability of GSTs, however, does not ensure that users can employ these technologies competently. For example, an over-reliance on navigation systems has resulted in people losing their way. In remote locations, such as California's Death Valley, the lack of "competent application" of GST can be a matter of life and death (Clark 2011).

To apply spatial thinking effectively, individuals require a spatial awareness or spatial literacy that does not necessarily develop as a consequence of using GST. If, as Ellul (1964) argues, technology is a sociological phenomenon, GST has generated social change, which in turn has driven political change. These changes have created a society awash in spatial data yet lacking the cognitive skills and the spatial "habit of mind" to use that data to solve problems, make decisions, or affect policy. As a result, society has begun to recognize the importance of a population who are competent spatial thinkers. Interest in teaching and learning spatial thinking is increasing, creating a challenge and an opportunity for geography education to lead in establishing a spatial thinking culture.

3.4.1 Technological Advances Driving Sociocultural and Political Change

The spatial revolution started thousands of years ago with maps. A map can be considered technology, or tool, that facilitates problem solving and decision making. However, prior to the invention of the printing press, it was impossible to distribute maps widely. Following the printing revolution, maps became more commonly available contributing to a cultural shift from a "manuscript culture" to a "print culture" (Finkelstein and McCleery 2002).

A second technological advancement occurred during World War I when aerial photography was combined with cartography to create and revise maps. This integration of technologies led to a cultural shift in how the surface of the earth was perceived—from a "view from below" to a "view from above." It demonstrated that geographic phenomena, both physical and social, are embedded in patterned spatial relationships that cannot be seen from ground level and led to the emergence of a new way of conceptualizing socio-spatial relationships. This new conceptualization resulted from the advancement of technology. Once introduced, it was manifested in society and policy through instructional programs such as France's use of "air-mindedness" to promote nationalism in schools and to understand long-term urban spatial stratification from aerial photographs of French cities (Haffner 2013).

The current turn in the spatial revolution is driven by the pervasive nature of GST. Maps that were printed, static, and clumsy are now digital, dynamic, interactive, and convenient. Maps and the spatial information derived from them are available everywhere, all of the time. Maps can be modified, created, and displayed by novices. This has led to a rethinking of maps and new understandings about their roles in spatial practices (Kitchin and Dodge 2007). A map can be created almost instantaneously by anyone to organize and display spatial information or phenomena (Edelson 2012). These developments have encouraged research about teaching and learning spatial thinking in order that individuals acquire the necessary cognitive skills to productively participate in a spatial culture inundated with GST.

Any educational emphasis on spatial thinking will be influenced by the support of government agencies and policy makers. Typically educational innovations are implemented only after societal evaluation, a process influenced by the endorsements of legislatures and policy makers. As governments standardize data formats and as public resources are dedicated to spatial learning and educational practice, educators have an increasing responsibility to include spatial thinking in their instruction.

3.4.2 Educational Applications

Driven by the social and political changes resulting from advances in GST, a *proto-spatial thinking culture* exists today. This culture is characterized by the widespread availability of GST but the limited spatial thinking ability of most people to use these technologies effectively. Spatial thinking is perhaps the most important factor that determines competency in spatially dependent disciplines such as geography and other STEM disciplines (Newcombe 2010; Wu and Shah 2004; Pallrand and Seeber 1984; Hsi et al. 1997; Kali and Orion 1996; Shea et al. 2001). Numerous studies have found a significant correlation between spatial thinking and success in spatially dependent tasks, performance in undergraduate science courses, and persistence in science careers (Anderson and Leinhardt 2002; Black 2005; Kali and Orion 1996; Keehner et al. 2006; Pallrand and Seeber 1984).

Researchers indicate that spatial thinking ability increases students' likelihood of pursuing a degree or career in STEM (Kheener et al. 2006; Shea et al. 2001) and that spatial thinking is malleable—it can be improved with training (Newcombe 2010). Many educational interventions for improving spatial thinking have been reported: (a) attention to the acquisition and use of spatial vocabulary (Bednarz and Bednarz 2008; Gentner 2007); (b) facilitating mental images through the use of gestures (Newcombe 2010); (c) use of analogies to identify similarities between un-like objects or phenomena (Loewenstein and Gentner 2005); (d) use of student sketches to represent conceptual models (McNeal et al. 2008); and (e) use of representations such as maps and GST (Uttal 2000; Anderson and Leinhardt 2002).

Classroom-based research has found that the use of GST, specifically GIS, improves knowledge of spatial concepts and the ability to think spatially. Lee and Bednarz (2009) found a positive relationship between the number of GST courses (e.g., GIS or Computer Cartography) completed by undergraduate students and their scores on a spatial-skills test. Even students who completed only one GIS course showed significant gains in spatial skills. Working with elementary students, Shin (2006) used qualitative analysis to demonstrate a positive relationship between the use of GST and students' learning and cognitive strategies.

3.4.3 Spatial Thinking in the Curriculum

This section asks, what is the interaction between acquiring content knowledge and spatial thinking? Three types of spatial thinking exist, depending on the context in which the spatial thinking takes place (NRC 2006). Walking to school or playing a team sport such as football—actions that are performed in space—require spatial thinking in a real-world context, referred to as the "geography of life spaces." The second type of spatial thinking, thinking about space, is typically employed when individuals use spatial thinking to learn how the world works. It is termed "the geography of physical spaces." The third type of spatial thinking, thinking with space, is abstract yet powerful. Educators often encourage students to "map" their understanding of relationships and concepts. We make lists, doodles, graphic organizers, diagrams, and graphs to explore data. Spatializing non-spatial data or using space as an organizing framework is an effective cognitive strategy. This third context results in a "geography of intellectual spaces."

The opportunity for students to learn "in space" varies greatly. Informal education such as scouting or sports can engage learners' abilities in wayfinding and in understanding location in space. Many other opportunities to learn about space exist, particularly in science courses such as geography, but the essence of spatial thinking, thinking about space, is rare in curricula we have examined. We argue for a concerted effort to introduce and institutionalize this perspective, and essential element of geography, into curriculum worldwide.

Previous research indicates that a spatial-thinking curricula must consider five issues. First is the importance of the individual learner. Spatial skills develop uniquely for different individuals. Sex, experience, age, culture, and education all play a role in the acquisition of key spatial thinking competencies. Second, context matters. Research confirms that expertise develops in specific contexts or disciplines and that transfer from one area to another is not automatic (National Research Council 1999). Thus, curricula should be developed across disciplines with spatial thinking in mind. Third, scale matters. Differences in large-scale and small-scale spatial thinking exist similar to differences in thinking in, about, and with space (Hegarty et al. 2006). Fourth, task analysis and alignment matter. It is essential that curricula include activities aligned to the three types of spatial thinking (*in* space, *about* space, or *with* space) in order to clarify the kinds of experiences that promote spatial skills and to understand the roles that individual differences play in spatial thinking. Finally, teaching matters. The findings indicate that spatial thinking can, and should, be learned by everyone (NRC 2006; Bednarz and Bednarz 2008). Thus, teacher preparation and professional development are key to improving spatial thinking (Jo and Bednarz 2014).

3.5 Assessment

In addition to questions concerning the conceptualization of spatial thinking and strategies to improve it, another issue involves the assessment of spatial thinking. Before researchers can detect changes, they must be able to measure an individual's spatial thinking ability. Spatial ability, a concept generally thought to be more specific than spatial thinking, has been the subject of research by practitioners from a wide range of disciplines.

In general, researchers agree that at least two spatial abilities exist, spatial visualization and spatial orientation. Tests are available to measure visualization—the ability to mentally represent and operate on visual stimuli—and orientation—the ability to picture spatially arrayed elements from different perspectives (e.g., Goldstein et al. 1990; Kail et al. 1979; Newcombe and Dubas 1992). These assessment instruments, however, leave many geographers and earth scientists dissatisfied because of their small scale and the restricted set of abilities they measure (Self et al. 1992; Golledge 1993; Montello et al. 1999; Lee and Bednarz 2009). Geographers and their colleagues have expressed a desire for an instrument that would assess what Golledge and Stimson (1997, p. 158) defined as spatial relations: "…abilities to recognize spatial distributions and spatial patterns, to connect locations, to associate and correlate spatially distributed phenomena, to comprehend and use spatial hierarchies, to regionalize, to orientate to real-world frames of reference, to imagine maps from verbal descriptions, to sketch map, to compare maps, and to overlay and dissolve maps."

The lack of test instruments for these abilities or skills leads researchers to create several "spatial analysis tests" (Kerski 2000; Marsh et al. 2007; Huynh and Sharpe 2013; Jo and Bednarz 2009). Many of these assessments were developed for relatively specific purposes, often based on specific curriculum elements (Huynh and Sharpe 2013) or to evaluate students' prerequisite knowledge (Jo and Bednarz 2009). Other tests were not examined for their validity and reliability. One recently developed assessment, the Spatial Thinking Abilities Test (STAT), was created using a recommended, five-step, test-development procedure (Lee and Bednarz 2012). Questions were created based on hypothesized spatial thinking components identified in the literature (Golledge et al. 2008; Gersmehl and Gersmehl 2006).

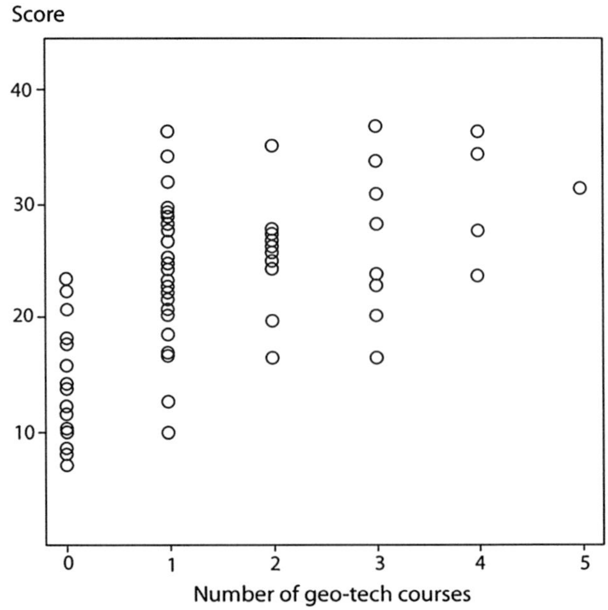

Fig. 3.1 Number of GIS courses completed by the subjects and their average scores in the spatial-skills test

The Spatial Thinking Abilities Test has been administered widely to a diverse set of subjects in many countries. It was used to measure subjects' mastery of the content and skills contained in the Association of American Geographers' *Teachers' Guide to Modern Geography*, a program to improve the ability of geography teachers to introduce spatial thinking into the classroom. STAT was also administered to undergraduate students to determine the effect of completing a geo-spatial technology course (i.e., GIS, cartography, or remote sensing) on their spatial thinking skills (Lee and Bednarz 2009). The pilot test revealed a positive correlation of 0.58 between the number of GIS or geo-technology courses completed and students' STAT scores (Fig. 3.1).

Following the pilot, 80 students enrolled in Computer Cartography, Introduction to GIS, and Economic Geography in a department of geography at a large public university completed both pre- and post-tests, before they began and after their semester-long courses (Table 3.1).

For additional applications of the STAT and a call for additional research on spatial thinking assessment, see Lee and Bednarz (2012) and Kim and Bednarz (2013).

Table 3.1 A comparison of pre- and post-test scores by group

		Pre-test		Post-test		
	N	Mean	S.D.	Mean	S.D.	Score difference
Control	35	11.171	5.046	11.71	4.773	0.542
Cartography only	18	12.972	5.829	14.11	4.629	1.138
GIS only	17	12.500	5.172	14.97	4.777	2.470*
Cartography and GIS						
Sequentially	7	17.571*	3.194	19.00	2.449	1.428
Concurrently	3	12.333	5.276	13.83	4.963	1.200

Students' average scores (Table 3.1 and Fig. 3.2) indicated that GIS students scored significantly higher on the post-test, although it should be noted that the number of students in each group is not large

*$p < 0.05$

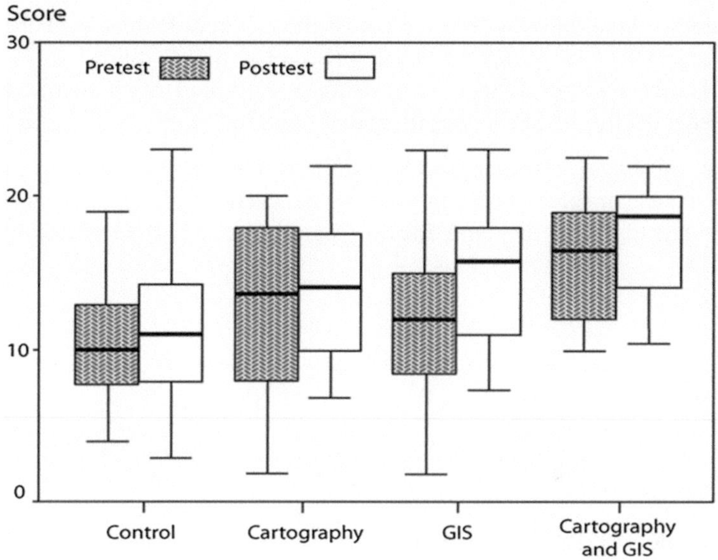

Fig. 3.2 A comparison of pre- and post-test scores by group

3.6 Conclusion

The chapter concludes with recommendations for further research. Technological changes, the spatial turn, and consequent social and political forces are producing a demand for a citizenry with the knowledge, skills, and practices of spatial thinking. The implications for geography educators are immense. People will need and want to know how to acquire, interpret, and contribute geographic information. The task increasingly is to prepare technologically enabled, spatially literate individuals. It will be a challenge (but also a considerable opportunity) to plan and to provide a level of understanding of spatial concepts and reasoning processes through geography (Bednarz et al. 2013). First, we may need to examine educational standards and their role in instruction in spatial thinking. In the United States, the National Geography Standards have been revised to embrace spatial thinking as a central mission. The term *spatial thinking* was inserted into the first Standard, *How to use maps and other geographic representations, GSTs, and spatial thinking to understand and communicate information.* Enhanced expectations regarding the use of GST to produce and interpret maps and solve spatial problems were included across all 18 Standards. The impact of this educational change remains to be researched.

A second area that remains unclear in a research sense is the relationships between spatial thinking and geography education, particularly at the intersection with GST. To focus and build capacity in geography education related to spatial thinking, four research questions are suggested to frame an agenda:

1. How does spatial thinking develop across individuals, settings, and time?
2. How does spatial thinking develop across the different realms of geography?
3. What supports or promotes the development of spatial thinking?
4. What is necessary to support the effective and broad implementation of the knowledge, skills, and practices of spatial thinking?

Finally, we suggest researchers carry out lines of research with select attributes, including the use of shared tasks, measurements and assessments such as the STAT to accumulate data on the core ideas, practices, and characteristics of spatial thinking.

References

Allen, G. L. (2003). Functional families of spatial abilities: Poor relations and rich prospects. *International Journal of Testing, 3*(3), 251–262.

Anderson, K. C., & Leinhardt, G. (2002). Maps as representations: Expert novice comparison of projection understanding. *Cognition and Instruction, 20*(3), 283–321.

Bednarz, R. S., & Bednarz, S. W. (2008). The importance of spatial thinking in an uncertain world. In D. Z. Sui (Ed.), *Geospatial technologies and homeland security* (pp. 315–330). Dordrecht: Springer Science.

Bednarz, S. W., Heffron, S., & Huynh, N. T. (2013). *A road map for 21st century geography education: Geography education research. Research document.* Geography Education Research Committee of the Road Map for 21st Century Geography Education Project. National Geographic Society. http://education.nationalgeographic.com/education/programs/road-map-project/?ar_a=1. Accessed 6 Oct 2014.

Black, A. A. (2005). Spatial ability and earth science conceptual understanding. *Journal of Geoscience Education, 53*(4), 402–414.

Clark, K. (2011). The GPS: A fatally misleading travel companion. *National Public Radio*. http://www.npr.org/2011/07/26/137646147/the-gps-a-fatally-misleading-travel-companion. Accessed 6 Oct 2014.

Downs, R. M., & Liben, L. S. (2001). The development of expertise in geography: A cognitive-developmental approach to geographic education. *Annals of the Association of American Geographers, 81*(2), 304–327.

Edelson, D. C. (2012). Unlocking the educational potential of citizen science. Resource document. *ArcNews*. http://www.esri.com/news/arcnews/spring12articles/unlocking-the-educational-potential-of-citizen-science.html. Accessed 6 Oct 2014.

Ellul, J. (1964). *The technological society*. New York: Vintage Books.

Finkelstein, A., & McCleery, D. (2002). *The book history reader*. London: Routledge.

Gauvain, M. (2008). Vygotsky's sociocultural theory. In M. M. Haith & J. B. Benson (Eds.), *Encyclopedia of infant and early childhood development* (pp. 404–413). Oxford: Elsevier.

Gentner, D. (2007). Spatial cognition in apes and humans. *Trends in Cognitive Sciences, 11*(5), 192–194.

Gersmehl, P. J., & Gersmehl, C. A. (2006). Wanted: A concise list of neurologically defensible and assessable spatial thinking skills. *Research in Geographic Education, 8*, 6–37.

Gersmehl, P. J., & Gersmehl, C. A. (2007). Spatial thinking by young children: Neurologic evidence for early development and "educability". *Journal of Geography, 106*(5), 181–191.

Goldstein, D., Haldane, D., & Mitchell, C. (1990). Sex differences in visual–spatial ability: The role of performance factors. *Memory & Cognition, 18*(5), 546–550.

Golledge, R. G. (1993). Geographical perspectives on spatial cognition. In T. Gärling & R. G. Golledge (Eds.), *Behavior and environment: Psychological and geographical approaches* (pp. 16–46). Amsterdam: Elsevier Science.

Golledge, R. G. (2002). The nature of geographic knowledge. *Annals of the Association of American Geographers, 92*(1), 1–14.

Golledge, R., & Stimson, R. (1997). *Spatial behavior: A geographic perspective*. New York: The Guilford Press.

Golledge, R. G., Dougherty, V., & Bell, S. (1993). *Survey versus route-based wayfinding in unfamiliar environments*. University of California Transportation Center, University of California Berkeley. http://uctc.net/research/papers/214.pdf. Accessed 6 Oct 2014.

Golledge, R. G., Marsh, M., & Battersby, S. (2008). Matching geospatial concepts with geographic educational needs. *Geographical Research, 46*(1), 85–98.

Goodchild, M. F., & Janelle, D. G. (2010). Toward critical spatial thinking in the social sciences and humanities. *GeoJournal, 75*(1), 3–13.

Haffner, J. (2013). *The view from above: The science of social space*. Boston: MIT Press.

Hegarty, M., Montello, D. R., Richardson, A. E., Ishikawa, T., & Lovelace, K. (2006). Spatial abilities at different scales: Individual differences in aptitude-test performance and spatial-layout learning. *Intelligence, 34*, 151–176.

Hsi, S., Linn, M., & Bell, J. (1997). The role of spatial reasoning in engineering and the design of spatial instruction. *Journal of Engineering Education, 86*(2), 151–158.

Huynh, N., & Sharpe, B. (2013). An assessment instrument to measure geospatial thinking expertise. *Journal of Geography, 112*(1), 3–17.

Janelle, D. G., & Goodchild, M. F. (2009). Location across disciplines: Reflections on the CSISS experience. In D. Z. Sui (Ed.), *Geospatial technology and the role of location in science* (pp. 15–29). Dordrecht: Springer Science.

Jo, I., & Bednarz, S. W. (2009). Evaluating geography textbook questions from a spatial perspective: Using concepts of space, tools of representation, and cognitive processes to evaluate spatiality. *Journal of Geography, 108*, 4–13.

Jo, I., & Bednarz, S. W. (2014). Developing pre-service teachers' pedagogical content knowledge for teaching spatial thinking through geography. *Journal of Geography in Higher Education, 38*(2), 301–313.

Kail, R., Carter, R., & Pellegrino, J. (1979). The locus of sex differences in spatial ability. *Perception & Psychophysics, 26*(3), 182–186.

Kali, Y., & Orion, N. (1996). Spatial abilities of high-school students in the perception of geologic structures. *Journal of Research in Science Teaching, 33*, 369–391.

Keehner, M., Lipa, Y., Montello, D., Tendick, F., & Hegarty, M. (2006). Learning a spatial skill for surgery: How the contributions of ability change with practice. *Applied Cognitive Psychology, 20*, 487–503.

Kerski, J. J. (2000). *The implementation and effectiveness of GIS technology and methods in secondary education*. Unpublished doctoral dissertation, University of Colorado, Boulder.

Kim, M., & Bednarz, R. S. (2013). Development of critical spatial thinking through GIS learning. *Journal of Geography in Higher Education, 37*(3), 350–366.

Kim, M., Bednarz, R. S., & Kim, J. (2012). The ability of young children to use spatial representations. *International Research in Geographical and Environmental Education, 21*(3), 261–277.

Kitchin, R., & Dodge, M. (2007). Rethinking maps. *Progress in Human Geography, 31*(3), 1–14.

Kompridis, N. (2010). So we need something else for reason to mean. *International Journal of Philosophical Studies, 8*(3), 271–295.

Kuipers, B. (1978). Modeling spatial knowledge. *Cognitive Science, 2*, 129–153.

Lee, J., & Bednarz, R. S. (2009). Effect of GIS learning on spatial thinking. *Journal of Geography in Higher Education, 33*(2), 183–198.

Lee, J., & Bednarz, R. S. (2012). Components of spatial thinking: Evidence from a spatial thinking ability test. *Journal of Geography, 111*(1), 15–26.

Loewenstein, J., & Gentner, D. (2005). Relational language and the development of relational mapping. *Cognitive Psychology, 50*, 315–353.

MacEachren, A. M. (1992). Application of environmental learning to spatial knowledge acquisition from maps. *Annals of the Association of American Geographers, 82*(2), 245–274.

Marsh, M., Golledge, R., & Battersby, S. (2007). Geospatial concept understanding and recognition in G6–college students: A preliminary argument for minimal GIS. *Annals of the Association of American Geographers, 97*(4), 696–712.

McNeal, K. S., Miller, H. R., & Herbert, B. E. (2008). The effect of using inquiry and multiple representations on introductory geology students' conceptual model development of coastal eutrophication. *Journal of Geoscience Education, 56*(3), 201–211.

Montello, D. R. (1998). A new framework for understanding the acquisition of spatial knowledge in large-scale environments. In M. J. Egenhofer & R. G. Golledge (Eds.), *Spatial and temporal reasoning in geographic information systems* (pp. 143–154). New York: Oxford University Press. http://www.geog.ucsb.edu/~montello/pubs/microgenesis.pdf. Accessed 6 Oct 2014.

Montello, D. R., Lovelace, L. L., Golledge, R. G., & Self, C. M. (1999). Sex-related differences and similarities in geographic and environmental spatial abilities. *Annals of the Association of American Geographers, 89*, 515–534.

National Research Council. (1999). *How people learn: Brain, mind, experience, and school*. Washington, DC: The National Academies Press.

National Research Council. (2006). *Learning to think spatially: GIS as a support system in the K-12 curriculum*. Washington, DC: The National Academies Press.

Newcombe, N. S. (2000). So, at last we can begin. *Developmental Science, 3*(3), 276–278.

Newcombe, N. S. (2010). Picture this: Increasing math and science learning by improving spatial thinking. *American Educator, 34*(2), 29–43.

Newcombe, N. S., & Dubas, J. S. (1992). A longitudinal study of predictors of spatial ability in adolescent females. *Child Development, 63*, 37–46.

Newcombe, N. S., & Huttenlocher, J. (2003). *Making space: The development of spatial representation and reasoning*. Cambridge, MA: MIT Press.

Newcombe, N. S., & Learmonth, A. E. (2005). Development of spatial competence. In P. Shah & A. Miyake (Eds.), *The Cambridge handbook of visuospatial thinking* (pp. 213–256). New York: Cambridge University Press.

Pallrand, G. J., & Seeber, F. (1984). Spatial abilities and achievement in introductory physics. *Journal of Research in Science Teaching, 21*(5), 507–516.

Piaget, J., & Inhelder, B. (1948). *A child's conception of space*. English edition: Langdon, F. J. & Lunzer, J. L. (1967) New York: Norton.

Self, C. M., Gopal, S., Golledge, R. G., & Fenstermaker, S. (1992). Gender-related differences in spatial abilities. *Progress in Human Geography, 16*, 315–342.

Shea, D. L., Lubinksi, D., & Benbow, C. P. (2001). Importance of assessing spatial ability in intellectually talented young adolescents: A 20-year longitudinal study. *Journal of Educational Psychology, 93*(3), 604–614.

Shelton, A. L., & McNamara, T. P. (2004). Orientation and perspective dependence in route and survey learning. *Journal of Experimental Psychology, 30*(1), 158–170.

Shin, E. (2006). Using geographic information systems (GIS) to improve fourth graders' geographic content knowledge and map skills. *Journal of Geography, 105*(3), 109–120.

Thommen, E., Avelar, S., Sapin, V. Z., Perrenoud, S., & Malatesta, D. (2010). Mapping the journey from home to school: A study on children's representation of space. *International Research in Geographical and Environmental Education, 19*(3), 191–205.

Uttal, D. H. (2000). Seeing the big picture: Map use and development of spatial cognition. *Developmental Science, 3*(3), 247–264.

Wu, H., & Shaw, P. (2004). Exploring visuospatial thinking in chemistry learning. *Science Education, 88*, 465–492.

Chapter 4
Education for Spatial Citizenship

Thomas Jekel, Inga Gryl, and Uwe Schulze

Abstract This chapter deals with approaches to use geoinformation (GI) as used in everyday settings. It first explores the concept of Spatial Citizenship, along an example taken from the holiday crowd that is easily translated to a host of other interest groups using space as symbolic means to exert their interests. It then looks into the role that digital GI may play in that process, and fields of competences needed to use GI competitively for active / activist citizenship. Theoretical foundations of a coherent concept of Spatial Citizenship are discussed, as is the reception of the approach by the scientific community not involved in the original conception of the Spatial Citizenship approach. The second part of the paper is devoted to the development of a more formalized set of competences as well as a curriculum that should enable in-service teachers to teach their classes in secondary schools along the line of the Spatial Citizenship approach. The contribution finally gives an outline of the materials developed within a European Union Comenius project.

Keywords Geomedia • Citizenship education • Teacher training

4.1 Introduction

This chapter deals with approaches to use geoinformation (GI) as used in everyday settings. It first explores the concept of Spatial Citizenship, along an example taken from the holiday crowd that is easily translated to a host of other interest groups using space as symbolic means to exert their interests. It then looks into the role that

T. Jekel (✉)
Department of Geography and Geology, University of Salzburg, Salzburg, Austria
e-mail: thomas.jekel@sbg.ac.at

I. Gryl
Department of Geography, University of Duisburg-Essen, Essen, Germany
e-mail: inga.gryl@uni-due.de

U. Schulze
Department of Human Geography, Goethe University Frankfurt, Frankfurt, Germany
e-mail: uschulze@geo.uni-frankfurt.de

O. Muñiz Solari et al. (eds.), *Geospatial Technologies and Geography Education in a Changing World*, Advances in Geographical and Environmental Sciences,
DOI 10.1007/978-4-431-55519-3_4

digital GI may play in that process, and fields of competences needed to use GI competitively for active / activist citizenship. Theoretical foundations of a coherent concept of Spatial Citizenship are discussed, as is the reception of the approach by the scientific community not involved in the original conception of the Spatial Citizenship approach. The second part of the paper is devoted to the development of a more formalized set of competences as well as a curriculum that should enable in-service teachers to teach their classes in secondary schools along the line of the Spatial Citizenship approach. The contribution finally gives an outline of the materials developed within a European Union Comenius project.

4.1.1 Space and Citizenship

The ability to navigate society through being able to control specific dimensions of space is in no way new. Probably since the first world maps (Babylonian, Ptolemy), people would know about the importance of the whereabouts of places. To participate in society, we need systems to order things and phenomena in space, systems of representation of spatial phenomena, and systems of communicating and hiding meanings attached to space.

Let us look at an example that might look a little out of the way. Figure 4.1 shows a non-euclidic mapping of the Austrian ski area Hochfügen-Hochzillertal provided by the local tourism industry. At first glance, this image allows to recollect the whereabouts of lifts and cable cars of ski runs, and other routes as well as stations to replenish visitors' energy at various ski huts. It includes a classification of the difficulty of various runs through a color code. This feature clearly allows to relate your personal abilities to the terrain (e.g. steepness of the slope). In turn it

Fig. 4.1 An online tourist map of the Hochzillertal-Hochfügen ski area, Tyrol, Austria (Skiliftgesellschaft Hochfügen GmbH 2014)

allows to identify yourself as sporty or leisure skier. The maplike representation also clearly denotes areas you are not supposed to ski, for example, private lands, young afforestation, areas meant for hunting or retreat of deer in the winter season and so on. Even this representation includes modes of control or at least attempts to control specific actions. The upper left corner denotes a specific area denoted as Fun Park on the map and inhabited by snow boarders. In reality, this area has several artificial jumps and a half pipe, rather loud and specific music, will cater to a specific age range of boarders wearing a specific fashion that transports some kind of cool sub-culture feel. The area exudes a deep group identity shown off by symbols and a language and vocabulary not used in other places. At the same time, there is a clear sense as to who does not belong to that area or group – ordinary skiers who are perceived as way uncool and who are actively shown as much. At the same time, the group of boarders making this little place their own, are no social group in the sense of at least semi-permanent links and social contacts – they usually hail from all corners of central Europe and dissolve in the evening or as soon as their holiday ends, sharing a (spare time) life style.

The above depiction of life within and belonging to a boarder park allows for quite a few glimpses that we deem characteristic of today's citizenship and their connections to space:

- Citizenship is enacted through every day action within physical space. It includes acts of temporary appropriation of space.
- Citizenship is temporary and fluid, as 'inhabitants' disperse in time (in the evening or at the end of holidays)
- Citizenship is spatially referenced yet not physically or administratively bordered, but linked to everyday regionalization
- Citizenship relies on rules that are enforced by both symbolic as well as spatial means. Those rules may be discussed or changed, and are therefore subject evolution and revolution by both individuals and subgroups.
- Citizenship is embedded in (or fighting) both administrative regulations as well as the multi-million-Dollar business models of the tourism industry who carefully tends to and makes money out of self-perceived sub cultures.
- Citizenship is including a sense of belonging to and identification with a group through symbolic, spatial and ideological means.

You may of course translate these basic ideas of citizenship from the heights of the Austrian Alps to a wider sphere of social entities and movements. Think of the "Occupy movement" opposing current regulations of the finance industry who carries appropriation of space in its name, think of environmental or local initiatives – they all include a spatial reference and use it partly overt, partly implicit in their internal and external organization. The next question then is: How do modern media in general and geomedia in particular interact with this fluid conception of citizenship? And are there open ends secondary education has to address to enable students to successfully participate in multiple citizenships?

4.1.2 Citizenship and Digital Spatial Information

Let us return to the original example. In reality, the above map of the skiing area is of course an interactive web tool. It provides a few added features indicating that citizenship may have changed with the advent of daily GI use.

First, underlying various infrastructure resources like lifts and ski runs is additional information covering capacities, opening hours and technical information of the area. Second, the skiing area also provides you with an app based on GPS Data that allows you to easily document and share your achievements on the mountain. As pointed out by Winkler et al. (2013) this not only allows you to help your self-esteem, but it also helps the cable car industry to actively collect data on your spatial behavior (e.g., regarding abilities, speed, feeding patterns, unlawful routes and so on). It is not the skiers only who appropriate their space, it is industry who appropriates the spatial data of the individual. Thus, the above-mentioned groups of skiers share their spatial footprint both along their abilities, as well as social norms they link into. The same as soon as we use our cell phones in everyday life, the data collection being enhanced by various apps.

It has to be said that there is nothing wrong with the sharing of spatial data. However, it has to be made sure that (a) the collectors adhere to certain rules of use, as well as (b) the providers – the individual skier, and the individual mobile users in general – are aware of the fact and the ways in which the data might be used.

The use of geo-media also includes the active communication with spatial means, like geo-referenced text messages denoting location, panoramio, or place specific forums. These are used to document personal exploits to others, to document you are fitting in a specific lifestyle, and to communicate which places may be especially cool regarding that lifestyle.

The technical competences needed to work the above tools are minimal. More competences are needed regarding communication – including group-specific language – and visualization, i.e., how to produce *that* image of an ultra cool jump.

To sum up, the production of and communication with geomedia has become so easy that it has become ubiquitous and extensively used in everyday lives. In many cases, users do not even know they are using or producing geomedia. This is the situation the concept of education for Spatial Citizenship tries to address. The concept re-centers the use of geomedia in secondary education from a useful tool of science education to a tool to participate in society, and a tool to actively change. To support these aims, a very different approach to GI in secondary education is needed, including specific competence models for teachers and curricula for teacher training.

4.1.3 A Basic Concept of Spatial Citizenship

A spatial citizen should be able to interpret and critically reflect on spatial information, communicate with the assistance of maps and other spatial representations, and express location-specific opinions using geomedia.

When the term Spatial Citizenship was first introduced (Jekel and Gryl 2010; Gryl et al. 2010), it was influenced by two major schools of theoretical tradition that can be rather closely linked together, Critical Cartography and Critical GIScience. Both underline the function and power of maps (i.e., Harley 1989; Wood 1993). Transferring both into an educational setting informed by social geography and new cultural geography, the first argument therefore was:

Children make geographies
+ Geographies need powerful visualizations
+ Maps are the most powerful visualizations
= Children make Maps

(Jekel 2008).

The argument was close to a second strand of thought that developed through the reaction of GIScience to criticism mounted mainly by social science aware GIScientists (Pickles 1995; Schuurman 2000): They developed uses for Geographic Information Systems for participation (Elwood 2006). While these systems were still heavily relying on expert knowledge, they made it a point to include lay knowledge in public decision making. The decision making, however was still largely oriented at formal spatial planning decisions and the spatial approach clearly linked to boundaries of administrative entities.

From these two entry points, the original model of competences to be acquired by students included three areas to be touched upon during secondary education (see Gryl and Jekel 2012):

- A reduced set of technical competences regarding geoinformatics
- The ability to reflect on others spatial representations, as well as to reflect on one's own geo-media use
- The ability to use spatial representations in communication and participation.

4.2 Theoretical Background

This first set of competences owed much to the original set of theoretical foundations used. However, both the technological development easing technical requirements for the user as well as further research led to a widened understanding of needs for an education of Spatial Citizenship that has been more implicit in the original model.

Concepts of Space To make a concept of Spatial Citizenship viable, concepts of space have to be re-thought in the context of GI-based education. Differing from the spatial thinking approach based on absolute space as proposed by the National Research Council (2006, see Gryl and Jekel 2012), the political sphere clearly demands relational approaches to space. Relational approaches to space accept space as socially constructed and intentionally used in political (and everyday) communication and action (Massey 1998; Kitchin and Dodge 2007). Acceptance of the constructedness of space allows for both the analysis of political processes as well as for the envisioning of alternative meanings of space (Jekel 2007). Drawing on our example of the skiing area, the groups of snowboarder and skiers clearly carry rather different connotations of that space, and represent it accordingly. Space is a concept to order the world according to one's interest. At the same time, the technological background of the tools used are clearly reliant on the absolute concepts of space. Education for Spatial Citizenship therefore has to include relevant concepts of space as used in everyday action as well as scientific and technological discourses. A minimum setup would include absolute, perceived and relational space.

Critical Cartography/Critical GIScience The acceptance of the social construction of space as a basis to analyze human environment interaction also opens up to the debates on the functioning and power of spatial representations, i.e. maps and geographic information systems. Debates here have started with Brian Harley's (1989) deconstruction of maps in a historical perspective pointing to the fact that maps do also act as a representation of social relations, and therefore, power and control. Later work in the GIS domain has also focused on the analytical prowess of GIS that was closely linked to the spatial approach. This critique is centered around the mechanistic spatial approach as well as the exclusion of the ordinary citizen from the expert systems (Pickles 1995; Schuurman 2000, 2009; Harris and Harrower 2005). As pointed out by Schuurman (2009), Critical GIScience contributed to the development of a few research areas within GIScience as a reaction, including investigations on different ontologies, counter mapping, and Public Participation GIS (PPGIS). However, Critical GIScience may also be translated to educational use at secondary level, as it informs reflexive practice regarding spatial representations (Gryl 2009; Jekel 2008). However, little research and school examples have been provided so far looking into the now ubiquitous GI use by lay people.

Counter Mapping and Participatory GIS One of the main research areas resulting from early critiques of GIS was the development of public participation GIS, specifically for the area of spatial planning. Early examples were clearly influenced by the spatial approach again and did mainly allow local stakeholders to place different weights to specific factors (layers) that were present in a spatialized decision making process. Counter Mapping, on the other hand, tried to provide lay users to produce their own maps with the intent to further their interests, for example through the possibility to document land possession. Both approaches have been further developed through the geoweb in recent years and are now technically proficient to allow students at secondary school to contribute to maps under the collaborative mapping tag (see, for example, Vogler et al. 2012).

Citizenship Education Traditionally, citizenship education should prepare students for 'dutiful' citizenship (Bennett et al. 2009), allowing them to function within a given set of societal rules. Accordingly, citizenship education was often seen as a tool for nation-building on both national and international (e.g., the

European Unions) scale. As mentioned above, a widened, fluid, and possibly rule changing concept of citizenship is employed here, including emancipatory aims and concepts of activism (Elwood and Mitchell 2013).

4.2.1 Basic Competence Dimensions for Spatial Citizenship

The basic competence dimension for citizenship education need some distinction in (a) competences needed by students in secondary education, and (b) competences needed by teachers to help students acquire the basic competences. Here, we concentrate on competences students should acquire, while teacher training needs are dealt with in Sect. 4.3 of this chapter. The guiding question therefore is which competences are needed by the ordinary citizen to successfully participate in society and decision making. As denoted in Fig. 4.2, we consider three main areas that we detail below:

Technology and Methodology to Handle Geomedia Participating in the process of spatial communication requires technological competences, including the ability to actively handle geomedia of the period in question. As technologies develop, requirements vary, and generally, decrease. From the technological viewpoint, traditional map reading and drawing skills need to be revisited and adjusted: New technology allows for broadening of possibilities for the lay user, ranging from consumption to production of competitive geomedia in the framework given by the GI-tool. The spatial citizen, as opposed to spatial analysts and spatial information system managers, is considered to have five competences within the field of technology/ methodology (see also Strobl 2008, p. 136):

- Consumption: map reading, orientation and navigation
- Analyzing: using existing functionality to answer simple questions and fulfill single-step analytical tasks, developing hypotheses from spatial representations.

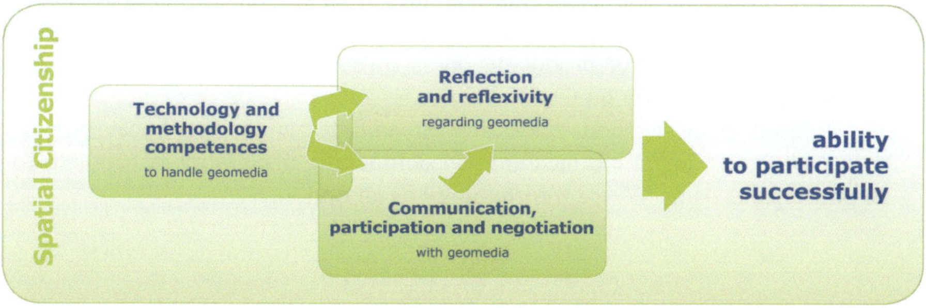

Fig. 4.2 Basic dimensions of education for spatial citizenship (Adapted from Gryl and Jekel 2012)

– Prosumption (Strobl 2008): changing data selection and visualization within a certain degree of freedom and participating by labeling, marking and commenting
– Producing: contribute one's own data and ideas
– Social networking: being able to use decision negotiation instruments on the web 2.0

Technical competences however, cannot be considered the final aim of an education for active Spatial Citizenship. They must be considered as preconditions for a reflected appropriation of space and active participation in society.

Reflection and Reflexivity Regarding Geomedia This component refers to the "consumption aspect" of handling geomedia. The classical consumption skills of map reading have to be extended by through deconstruction, conscious hypotheses production and envisioning space. It is therefore necessary to shift from decoding absolute spatial representations to reading them as the representations of the multiplicity of intentionally constructed relational spaces (Kitchin and Dodge 2007). This process can be termed reflection and reflexivity of spatial representations. Both require willingness for uncertainty and an attitude for reflectiveness as well.

– Reflection: knowing about the naturalization of spaces in geomedia and apply this knowledge to a certain spatial representations, comparing information with pre-knowledge and other sources, identifying hidden and missing information, thinking of alternative attachment of meanings and spatial scenarios.
– Reflexivity: knowing about own hypothesis generation with geomedia, reflecting own consumption processes, being aware of the own construction of spatial scenarios based on medium, preconditions, and own interest, developing alternatives, and deciding for acceptance of spatial scenarios or promoting alternatives.

Communication, Participation and Negotiation With Geomedia In addition to technological competences, Spatial Citizenship calls for competences for active communication and participation strategies.

– Expression: finding a way to convincingly communicate constructions of meanings and alternative, non-mainstream spatial scenarios, using GI.
– Communication: sharing ideas and meanings with the intention that communication partners adopt them, either using institutionalized online and offline communication paths, or producing one's own using the power of emerging communities, especially on the Web 2.0.
– Negotiation: engaging and discussing in an interactive, non-linear process, trying to reach compatible meanings in democratic negotiation acceptable for all participants using Web 2.0 technology as an option.

4.2.2 *Discussion: Reception of the Spatial Citizenship Approach*

Since its emergence, the Spatial Citizenship approach has already found significant reception. Besides the construction and field-testing of practical learning environments (e.g. Jekel et al. forthcoming; Kanwischer and Gryl 2012), several authors enhance and refine the theoretical foundations and seek for further fields of application.

Kanwischer et al. (2012; cf. Gryl, Schulze and Kanwischer 2013) supplement the initial concept with extracts of competences catalogues from different disciplines to produce a detailed spatial citizenship competence model and curriculum. With this, Spatial Citizenship is linked to the competence debate in secondary education on an international level.

Carlos and Gryl (2013) focus on another theoretical refinement of Spatial Citizenship by comparing it with the approach of Critical Thinking. Both Critical Thinking and the ancestor of Spatial Citizenship, Critical GIScience, share the term 'critical', and both approaches are close to concepts of citizenship (education). In contrast to Spatial Citizenship, Critical Thinking is an already influential approach, being received worldwide, and beyond the subject of geography. Despite some non-congruencies between both approaches, Spatial Citizenship can benefits from Spatial Thinking concepts such as rationality, moral, and creativity.

Elwood and Mitchell (2013) identify another significant aspect to enhance the Spatial Citizenship approach. Political action within Spatial Citizenship is initially described with the term 'participation', which, however, pretends equality among the different stakeholders that is not existent in society. The authors argue, that even if lay users may produce competitive geomedia with simple mapping tools and communicate counter narrations, direct confrontation is seldom possible. Thus, spatial citizens will not be able to realize strategic practices – the practices of the powerful that construct spaces permanently – but will be able to gain profit from tactical practices (cf. De Certeau 1984). Geomedia does not directly support political action, but may first and foremost boost political formation, in terms of "(1) the formation of political subjects, (2) the formation of interpretive frames that can mobilize these subjects for action, and (3) the formation of shared knowledge through collaborative cartographies" (Elwood and Mitchell 2013, p. 280). Therefore, tactical practices, in other words, visual spatial tactics have to be understood as condition for Spatial Citizenship, as they "can foster political subject formation, collaborative, interactions that generate shared knowledge, and critical insights that mobilize these subjects for engagement" (Elwood and Mitchell 2013, p. 288).

Kanwischer and Quennet (2012) open up a more application-oriented focus and stress the potential of Spatial Citizenship to play an important role in the ongoing process of the introduction of e-governance in developing countries in Africa. It may help to connect information and communication technologies to spatial decision making and simultaneously support the citizens' maturity.

Vogler et al. (2012) suggest the concept of Spatially Enabled Learning, extending the Spatial Citizenship's focus on the spatial domain to further learning processes in several subjects beyond geography. The authors stress the argument that learning with additional spatial orientation patterns displayed by geo-visualization may lead to deeper cognitive elaboration (cf. Paivio 1986). This approach does not only go beyond the approach of Learning to think Spatially by including the principles of Spatial Citizenship such as the social construction of spaces, it clearly turns this idea on its head by suggesting that geomedia might become a support system for learning in general (Vogler et al. 2012).

All these receptions of the original concept suggest that Spatial Citizenship may be a very useful basic concept for education in the spatial and political sphere that needs further clarification and empirical work. A first approach has been done while providing relevant curricula and materials for teacher training.

4.3 A Curriculum: Teacher Training for Spatial Citizenship Education

In order to make Spatial Citizenship tangible for teaching and learning in secondary school its theoretical contexts have to be operationalized in the form of a curriculum for teacher education and training. This task has been subject to the multilateral research project SPACIT in which various experts from fields such as geography, GIScience, education, and politics are engaged (cf. http://www.spatialcitizenship.org/).

Generally, a curriculum constitutes an obligatory foundation for the creation of certain learning activities. It provides the necessary pedagogic, didactic as well as technical specifications on teaching and learning such as the description of learning field(s) and competences, the learning outcomes and methods of assessment (cf. Cedefop 2010). In order to arrive at sound (pre-) structuring of an effective learning experience the curriculum development process should result in an appropriate configuration of the curricular content. Above all, that is, setting the scope and depth as well as the sequence and continuity of the learning content (Unwin et al. 2012; Painho and Curvelo 2012).

The development of the *SPACIT Curriculum for Teacher Education and Training included* the following aspects (cf. Schulze et al. 2014):

Competence Modelling For the definition of a coherent "Spatial Citizenship" learning construct, a competence model for Spatial Citizenship teacher education and training has been developed (cf. Kanwischer et al. 2012; Gryl et al. 2013). Integrating theoretical and conceptual findings from various neighboring domains of Spatial Citizenship, this model structures teachers' knowledge, skills and abilities within six major dimensions of (1) geomedia technology and methodology, (2) geomedia reflection, (3) geomedia communication, (4) the spatial domain, (5) the citizenship education domain, and (6) implementation strategies in school.

Formulation of Learning Outcomes The identification of Spatial Citizenship competences has been followed by the formulation of learning outcomes. They describe the goals of Spatial Citizenship learning processes in terms of general statements of "what a learner is expected to know, understand and be able to demonstrate after completion of learning experience" (Gonzalez and Wagenaar 2008; cf. also Kennedy et al. 2006). On the basis of the Revision of Blooms Taxonomy of Education Objectives by Anderson and Krathwohl (2001) SPACIT learning outcomes comprise different dimensions of knowledge as well as cognitive processes, and, therefore, can be associated with different learning activities (cf. Krathwohl 2002). In order to emphasize the dual function of the learning outcomes – teachers are 'learners' as well as 'professionals' in the field of Spatial Citizenship – the following phrase has been introduced:

Example: *At the end of the learning process the teacher should be able to create a learning environment to enable pupils to...*

Content Organization The SPACIT curriculum serves as the consistent basis to create local in-service teacher training opportunities across the European Higher Educational Areas (EHEA) contributing for lifelong learning purposes. Aiming at application in a variety of institutions the SPACIT curriculum demands on flexibility for the formulation, creation and the assessment of local teacher training settings and materials. Therefore, as well as due to the heterogeneity of the field of learners from different disciplines and professions the SPACIT curriculum has been designed as a multipath curriculum. This approach should allow for an individual handling of the learning content according to the specific needs and interests of engagement in particular competence areas of Spatial Citizenship.

Table 4.1 provides an overview of the four modules of the SPACIT in-service teacher training environment which assigns the different competence areas of Spatial Citizenship education to respective learning contents of the application-oriented use of digital geomedia.

Approaches of Learning, Teaching, and Assessment Spatial Citizenship education deals with mainstream technology. Therefore, fostering teachers' technical pedagogical content knowledge (cf. Mishra and Koehler 2006; Fargher 2006), approaches of teaching and learning for Spatial Citizenship means integrating a variety of digital tools ('hands on') as well as examples of geomedia based communication. On the one hand appropriate learning settings therefore must aim at different modes of self-directed online and offline face-to-face learning. Paying attention to the teachers' individual (fore) knowledge and demands those learning environments should allow learner-centered and active approaches of teaching and learning. On the other hand this also comprises appropriate forms of assessment in order to evaluate the outcomes of learning both the lecturer and the learner. Beside

Table 4.1 Description of the modules of the SPACIT in-service teacher training environment

Module: Concepts of Space	
Units	*Introduction to Spatial Conceptions; Construction of Space & Spatial Thinking; Action & Space*
Description	*This module aims to provide the learner a comprehensive engagement with absolute and relative spatial concepts and representations. Related to the mature appropriation of space this module focuses on the awareness of relational spatial conceptions and its consequences for societal power relations and (social) action within the physical environment. Relative concepts of space involve the social construction of spaces by the attachment of meaning to physical matter. Following the Spatial Thinking approach (NRC* 2006*) physical space is referred to with absolute concepts of space, tools of representation and processes of reasoning*
Module: Geomedia (GM) Technology and Methodology	
Units	*Introduction to GIS&T Domain; Basic Geomedia Applications; Advanced GM Applications*
Description	*This module aims to enable the learner for the creative and meaningful utilization of GM as powerful instruments of everyday social action. Therefore, related to the domain of Geographic Information Science & Technology (GIS&T) this module provides the technical as well as methodological knowledge and skills necessary for the mature handling of geospatial data in the form of GM information processing,* i.e. *consumption, production and prosumption of GM, analysis carried out using GM as well as aspects of technical communication in the form of social networking*
Module: Geomedia Communication and Reflection	
Units	*Introduction to GM Communication; Counter mapping & Alternative Spatial Visions; GM in Everyday Practices*
Description	*The aim of this module is to introduce the learner to the opportunities and challenges of how to express, communicate and critically interpret alternative spatial visions and constructions within GM. On the one hand this involves the creation of own spatial representation as regards content with the help of different GM like digital maps, texts, photographs* etc., *primarily basing on the tradition of counter mapping. On the other hand this module supports reflection concerning the consumption of GM and its influence on one's own and people's everyday action. Therefore traditional map reading skills are adjusted in order to create awareness: (a) of the translation between social and absolute space; (b) of GM as social constructions with limited representation of the world needed to be deconstructed; and (c) of the user's needs to be reflexive towards her/his own GM consumption*
Module: Concepts of Citizenship Education	
Units	*Introduction to Citizenship Education; Participation & Spatial Action; Negotiation & Decision Making with GM*
Description	The aim of this module is to introduce the learner to the concepts of citizenship education essential to understand the pivotal values for Spatial Citizenship. With special attention to the role of fluent institutions and (new media) communities as well as power relations in society the learner will be acquaint to the knowledge, skill and abilities necessary for (spatial) participation and decision making in society against the background of considering societal rules as fundamentally negotiable

Note: Each module is separated into three units of max. 6 h of workload each. Depending on the individual prior knowledge and competences within the particular learning fields, teachers are free to select the number of units to work through. For certification purposes two-thirds of all units have to be completed, at which at least one unit must be assigned from each module the application of elements of formative assessment, for instance, online self-assessment (e.g. quizzes), and interactive tasks (e.g. concept mapping), summative assessment takes place through a portfolio (e.g. online blog) for documentation and reflection purposes.

4.4 Conclusion

Both the curriculum and materials supporting it have been tested in a pilot course with 28 in service teachers in Iasi, Romania in July 2014. First results of the evaluation show that teachers were happy with the overall architecture as well as the new skills and perspectives Spatial Citizenship provides for geography education in general. At the same time, some of the coursework was found aiming to high, or needing better tutorials. Currently, the materials are finalized and will be made available free of cost. A strategy has been developed to include the curriculum and materials in several European universities, as well as in in-service teacher training.

In its current form, the Spatial Citizenship approach as developing geomedia competencies from every day, non-expert uses of geomedia has found some acceptance in secondary education. This is clearly a change from most geomedia uses in education that are centered on science and technology education. It is, however to be seen if the general concept can also be diversified, for example in terms of age and gender. First initiatives are currently taken to develop specific courses for senior citizens at the University of Salzburg, while gender-sensible restructuring of the original course should also take place in the foreseeable future.

Acknowledgement Work on this publication has been co-funded by the European commission under the: LLP-Comenius multilateral project SPACIT (517908-LLP-2011-1-AT-COMENIUS-CMP).

References

Anderson, L. W., & Krathwohl, D. R. (Eds.). (2001). *A taxonomy for learning, teaching, and assessing: A revision of Bloom's taxonomy of educational objectives* (Completeth ed.). New York: Longman.

Bennett, W. L., Wells, C., & Rank, A. (2009). Young citizens and civic learning: Two paradigms of citizenship in the digital age. *Citizenship Studies, 13*(2), 105–120.

Carlos, V., & Gryl, I. (2013). Where do critical thinking and spatial citizenship meet? Proposing a framework of intersections. In T. Jekel (Ed.), *GI_Forum 2013* (pp. 437–446). Berlin: Wichmann.

Cedefop (European Centre for the Development of Vocational Training). (2010). *Learning outcomes approaches in VET curricula: A comparative analysis of nine European countries* (Research Paper No 6). Luxembourg: Publications of the European Union. Available at www.cedefop.europa.eu/EN/Files/5506_en.pdf. Accessed 28 June 2014.

De Certeau, M. (1984). *The practice of everyday life*. Berkeley: University of California Press.

Elwood, S. (2006). Critical issues in participatory GIS: Deconstructions, reconstructions, and new research directions. *Transactions in GIS, 10*(5), 693–708.

Elwood, S., & Mitchell, K. (2013). Another politics is possible: Neogeographies, visual spatial tactics, and political formation. *Cartographica, 48*(4), 275–292.

Fargher, M. (2006). *Linking lessons learnt from the classroom with research findings on pedagogies with GIS*. Available at http://www.geography.org.uk/download/GA_PRSSFargher.doc. Accessed: 28 June 2014.

Gonzalez, J., & Wagenaar, R. (eds.) (2008). *Universities' contribution to the Bologna Process. An introduction*. Available at http://www.unideusto.org/tuningeu/images/stories/Publications/ENGLISH_BROCHURE_FOR_WEBSITE.pdf. Accessed: 28 June 2014.

Gryl, I. (2009). *Kartenlesekompetenz. Ein Beitrag zum konstruktivistischen Geographieunterricht*. Wien: Materialien zur Didaktik der Geographie und Wirtschaftskunde, Bd. 22.

Gryl, I., & Jekel, T. (2012). Re-centering GI in secondary education. Towards a spatial citizenship approach. *Cartographica, 47*(1), 18–28.

Gryl, I., Jekel, T., & Donert, K. (2010). GI & spatial citizenship. In T. Jekel, A. Koller, K. Donert, & R. Vogler (Eds.), *Learning with GI V* (pp. 2–11). Berlin/Offenbach: Wichmann.

Gryl, I., Schulze, U., & Kanwischer, D. (2013). Spatial citizenship. The concept of competence. In T. Jekel, A. Car, J. Strobl, & G. Griesebner (Eds.), *GI_forum 2013* (pp. 282–293). Berlin: Wichmann.

Harley, J. B. (1989). Deconstructing the map. *Cartographica, 26*(2), 1–20.

Harris, L., & Harrower, M. (2005). Critical interventions and lingering concerns: Critical cartography/GISci, social theory, and alternative possible futures. *ACME, 4*(1), 1–10.

Jekel, T. (2007). What you all want is GIS2.0. Collaborative GI based learning environments: Spatial planning and education. In A. Car (Ed.), *GI-crossroads @ GI-forum* (pp. 84–89). Heidelberg: Wichmann.

Jekel, T. (2008). Die Macht der Karten und die Macht der Kinder. Versuch einer Begründung des Lernens mit Geoinformation. In K. Dobler, T. Jekel, & H. Pichler (Eds.), *Kind: macht: raum* (pp. 62–75). Heidelberg: Wichmann.

Jekel, T., & Gryl, I. (2010). Geographie, geoinformation und politische bildung. In I. Juchler (Ed.), *Kompetenzen in der politischen bildung* (pp. 91–102). Schwalbach: Wochenschau Verlag.

Jekel, T., Gryl, I., Sanchez, E., Jouneau-Sion, C., & Lyon, J. (Eds.). (2014). *Learning and teaching with geomedia*. Newcastle upon Tyne: Cambridge Scholars.

Kanwischer, D., & Gryl, I. (2012). Der Einsatz von digitalen Karten und Globen zur Förderung der Argumentationskompetenz. In A. Budke (Ed.), *Diercke Kommunikation und Argumentation* (pp. 77–85). Braunschweig: Diercke.

Kanwischer, D., & Quennet, F. (2012). Distance education and spatial citizenship in Africa–Challenges and prospects. *Review of International Geographical Education Online, 2*(1), 95–117.

Kanwischer, D., Schulze, U., & Gryl, I. (2012). Spatial citizenship. Towards a curriculum. In T. Jekel, A. Car, J. Strobl, & G. Griesebner (Eds.), *GI_Forum 2012* (pp. 172–181). Berlin: Wichmann.

Kennedy, D., Hyland, A., & Norma, R. (2006). Writing and using learning outcomes: a practical guide. In E. Froment (Eds.), *EUA Bologna handbook*. Berlin: Raabe. Making Bologna work. Chapter C 3.4-1, Implementing Bologna in your institution. Available at: http://www.ehea-journal.eu/index.php?option=com_content&task=view&id=21&Itemid=148#b. Accessed 28 June 2014.

Kitchin, R., & Dodge, M. (2007). Rethinking maps. *Progress in Human Geography, 31*(3), 331–344.

Krathwohl, D. R. (2002). A revision of Bloom's taxonomy: An overview. *Theory Into Practice, 41* (4), 212–218.

Massey, D. (1998). *Power geometries and the politics of space-time*. Heidelberg: Department of Geography at University of Heidelberg.

Mishra, P., & Koehler, M. J. (2006). Technological pedagogical contents knowledge: A framework for teacher knowledge. *Teachers College Record, 108*(6), 1017–1054.

National Research Council. (2006). *Learning to think spatially. GIS as a support system in the K-12 curriculum*. Washington, DC: National Academies Press.

Painho, M., & Curvelo, P. (2012). Building dynamic, ontology-based alternative path for GIS&T curricula. In D. J. Unwin, K. E. Foote, N. J. Tate, & D. DiBiase (Eds.), *Teaching geographic information science and technology in higher education* (pp. 98–115). Chichester: Wiley.

Paivio, A. (1986). *Mental representations: A dual coding approach*. Ann Arbor/New York: University of Michigan/Oxford University Press.

Pickles, J. (Ed.). (1995). *Ground truth: The social implications of geographic information systems*. New York: The Guilford Press.

Schulze, U., Gryl, I., & Kanwischer, D. (2014). Spatial citizenship – Creating a curriculum for teacher education. In R. Vogler, A. Car, J. Strobl, & G. Griesebner (Eds.), *GI_Forum 2014* (pp. 230–241). Berlin: Offenbach.

Schuurman, N. (2000). Trouble in the heartland: GIS and its critics in the 1990s. *Progress in Human Geography, 24*(4), 569–590.

Schuurman, N. (2009). Critical GIS. In R. Kitchin & Thrift, N. (Eds.), *International encyclopedia of human geography* (Vol. 2, pp. 363–368). Oxford: Elsevier.

Skiliftgesellschaft Hochfügen GmbH. (2014). *Interaktiver Pistenplan*. Available at: http://hochfuegenski.com/content/pistenplan/interaktiver-pistenplan.html. Accessed 7 Nov 2014.

SPACIT (Spatial Citizenship). (2014). *Project website of the EU Comenius Project Spatial Citizenship*. Available at: http://www.spatialcitizenship.org/. Accessed 10 Oct 2014.

Strobl, J. (2008). Digital earth brainware. In J. Schiewe und U. Michel (Eds.), *Geoinformatics paves the highway to digital earth* (pp. 134–138). Osnabrück: gi-reports@igf.

Unwin, D. J., Foote, K. E., Tate, N. J., & DiBiase, D. (Eds.). (2012). *Teaching geographic information science and technology in higher education*. Chichester: Wiley.

Vogler, R., Hennig, S., Jekel, T., & Donert, K. (2012). Towards a concept of "spatially enabled learning". In T. Jekel, A. Car, J. Strobl, & G. Griesebner (Eds.), *GI_Forum 2012* (pp. 204–211). Berlin: Wichmann.

Winkler, B., Partl, A., Weilharter, K., & Maier, K. (2013). Einsatz der Applikation "Ski amadé Guide". *Unterricht. GW-Unterricht, 130*, 56–60.

Wood, D. (1993). *The power of maps*. London: Routledge.

Part II
Implementation of Geospatial Technologies in Formal Education and Informal Learning

Chapter 5
Integrating GIS and Other Geospatial Technologies in Middle Schools

Marsha Alibrandi and Donna Goldstein

Abstract This chapter describes theory, research and practice specific to middle school geospatial technology integration. From projects implemented in the late 1990s to the present, a variety of settings, project foci, and applications are reviewed. The unique pedagogical and curricular advantages and constraints on middle school geospatial technology integration situate geospatial thinking and learning in the young adolescent developmental context. The chapter presents evidence that middle school is an extremely age-appropriate opportunity for students to apply geospatial thinking and skills, generate new data, and develop a cognitive foundation for incorporating ever-expanding geospatial technologies and opportunities. From both reviewed and meta-analyses and specific applied research findings, we discuss successful projects from a variety of national settings and curricula. The adolescent developmental perspective is consistent with the theoretical statement that geographic practice is ontogenetic. This is congruent with the unique developmental advantages to introducing geospatial technologies across the middle school curriculum, beyond geography to integrated thematic settings. Integrating that practice with the developmentally appropriate activities of middle school students has yielded, in our findings and others' far-reaching results.

Keywords Middle school • Geospatial technologies • Curriculum integration • Adolescent development

5.1 Introduction

In this chapter, we focus more on the middle school aged learner, as compared to a larger number of discussions of teachers integrating geospatial technologies (GSTs). While the methodologies and results of the teaching are not ignored, they

M. Alibrandi (✉)
GISYES.org, Cummaquid, MA, USA
e-mail: marsh@cape.com

D. Goldstein
School District of Palm Beach County, West Palm Beach, FL, USA
e-mail: donna.goldstein@palmbeachschools.org

O. Muñiz Solari et al. (eds.), *Geospatial Technologies and Geography Education in a Changing World*, Advances in Geographical and Environmental Sciences,
DOI 10.1007/978-4-431-55519-3_5

are not our central focus. Our goal for the chapter is to take a constructivist approach by starting with the students of middle school and intermediate secondary school ages from a more developmental perspective. The unique timing of their age grades, their expanding social and physical landscape explorations, their proclivities and preferences for technologies, and for the ever-expanding landscapes of internet, social media, and other applications all impact how adolescents now come to know themselves and the world. This perspective is consistent with this volume's theoretical statement that geographic practice is ontogenetic. Therefore, integrating that practice with the developmentally appropriate activities of middle school students has yielded, in our own and others' findings, far-reaching results.

From studies of middle school-aged students using GSTs, we will discuss both cognitive and problem-solving skills as significant outcomes for the age group. These will be drawn from both measurable findings and case studies. As Barbaree Duke, a middle school Geographic Information System (GIS) educator and author has found, "GIS pushes students up Bloom's Taxonomy. . .addresses all learning styles" and that, for her gifted magnet students, "offers Type II and III Renzulli activities" (Ash-Duke 2005).[1]

5.2 Twenty-First Century Middle School Students

The broader spectrum of the "middle school" population requires consideration of the global context of "youth," a term used by UNESCO in aggregating research on children in international contexts. Youth trends are included to both expand upon the context of "middle school" to include trends among highly transient youth such as immigrants, refugees, trafficked youth, increasing and/or declining youth demographics, and other influences of a geographic nature upon this important segment of the population.

> Young people move within their home countries as internal migrants, or beyond their borders as international migrants. The report focuses largely on the phenomena of international migration which increasingly has a significant impact on the origin, transit and destination countries and communities. The consequences are complex, context-specific and subject to change over time. They may be influenced by factors such as the type of

[1] Renzulli's Type II Enrichment is composed of: (a) creative thinking and problem solving, critical thinking and affective processes; (b) a wide variety of learning-how-to-learn skills; (c) skills in the use of advanced-level reference materials, and; (d) written, oral and visual communication skills. Type III Investigations are designed to: (a) apply interests, knowledge, creative ideas and task commitment to a self-selected problem or area of study; (b) acquire advanced-level understanding of the knowledge (content) and methodology (process) used within specific areas of disciplines, artistic areas of expression, and interdisciplinary studies; (c) develop authentic products that are primarily directed toward bringing about a desired impact upon a specified audience; (d) develop self-directed skills in the areas of planning, organization, resource utilization, time management, decision-making and self-evaluation, and: (e) develop task commitment, self-confidence, and feelings of creative accomplishment (Renzulli and Reis, in Callahan and Herthberg-Davis 2013).

> migration, migrant category, national migration policies, and programmatic interventions that are in place in a particular locale. (UN World Youth Report 2013)

Although we may not assume that all adolescents are participating in online or telecommunications networks, due to economic, geographic and other constraints, we acknowledge the likelihood of their eventual participation with those and GST. With cell phone Global Positioning System (GPS), millions more users are now routinely accessing geospatial information on a regular basis. Interactive GPS and GIS maps are seen in cars, public transport, on TV news, and with adolescents more likely to see satellite imagery, their world views are shaped by these representational media (Downs and Stea 1977; Gould and White 1974).

Further, to situate adolescents in a global context, the awareness that the youth of 2020 will have of their counterparts across the planet will be unprecedented. As well, the mobility of youth in the twenty-first century is substantially different than during the previous century. This becomes more relevant when we consider recent findings regarding the effects of geographic and GST integration in middle school immigrant and ESL/ELL students (Hinde et al. 2011; Goldstein and Alibrandi 2013).

5.3 Pedagogical and Theoretical Frames

Consistent with our previous work in the area of integrating GIS and other GST in middle schools, we use the construct of "middle school," with its intent and focus on integrated curriculum. The "middle school" concept of curriculum integration (Beane 1995), focuses on *thematic studies that cross disciplinary lines*. This parallels the newest areas of scientific research that incorporate multiple data sets and sources, uniting them in geographic space. The entire field of climate change research would not have been possible before the advent of GST. Therefore, GST should have a central role in the integrated middle school curricula, given their capacity to integrate data from multiple sources and disciplines. Because this focus is unique to an early adolescent age-grade, we also include relevant studies from "intermediate" and "secondary schools" found in international settings.

The concept of the middle school integrated curriculum poses a broader and more integrated scope for GST integration since, as we and others have previously written, the integration need not be bound or constrained to a specific discipline (Hinde 2005; Alibrandi et al. 1999; Hinde et al. 2007; Goldstein and Alibrandi 2013). For that reason, we continue to frame a discussion of GST integration from a theoretical frame based in constructivist and Multiple Intelligences theories. From a social constructivist perspective, all learning is constructed in a social context and is built upon students' prior knowledge. Thus, it is essentially a scaffolded "bottom-up" process.

We consider the adolescent's experience as that of expanding landscapes, with greater responsibility and freedom of movement, although in some cultures, these are bounded by gender roles. In their way-finding activities, adolescents leave the previous bounds of home to attend school, find sports fields, go to markets as either buyers or sellers, and explore with their peer groups (Wigglesworth 2003). In these fundamental activities, adolescents are using spatial skills to make connections that are essential to intellectual growth, as they connect previously unconnected parts of their surroundings, get lost and found, learn new routes, and most importantly, find ways to solve spatial problems to save time and travel, locate resources and participate in community life. In rural Peru, for example, when the itinerant teacher arrives, a bell is rung, and students may walk several miles to attend. In this way children and adolescent siblings naturally construct new knowledge, and the basis of spatial skill development that is critical to survival; not just the geographic learning that is valued in educational settings. These experiences set the cognitive stage for adolescents' natural developmental uptake of problem, project and place-based content and skill learning. The very nature of GST to integrate a variety of data complements adolescents' developmental need for connection and synthesis. Jay Giedds (2004) MRI findings of adolescents decreasing gray matter and increasing white matter primes middle level students to connect problem and place-based learning with both their prior knowledge and their cohort, making GST integration not just relevant, but developmentally appropriate.

Incorporating Gardner's (1985, 2011) Multiple Intelligences (MI) theory is also consistent with our previous research reporting (Alibrandi 2003; Goldstein 2010). Gardner identifies "spatial intelligence" as one of the multiple intelligences. In introducing his theory, Gardner used particular criteria in order for a group of skills, mastery and competencies to be considered as "intelligence." Gardner included among these criteria,

- an identifiable core set of operations – basic kinds of information-processing operations or mechanisms that deal with one specific kind of input
- a distinctive developmental history, along with a definite set of "end-state" performances
- an evolutionary history and evolutionary plausibility (1985, pp. 62–66)

More specific to "middle school" or integrated curriculum, is Gardner's extension of the nature of *pattern recognition*, to apply patterns instantiated through one intelligence (for example, spatial intelligence) across others, such as linguistic intelligence. But there are several less-emphasized features of MI theory. First, it is a theory of intelligences distributed among the members of a social species, and is therefore phylogenetic, meaning that humans depend on one another for different types of intelligence, and that this interdependence is inherent in our DNA as a social species. While few teachers are exposed to these elements of MI theory, they have embraced the theory in their teaching practice so much that it is ubiquitous in educational professional development and even plan books.

We have found that while geospatial skills, knowledge and performance may occur on one domain (in our case, a Social Studies GIS elective) they are also transferred knowledge that impacts other domains as demonstrated on standardized tests. Therefore, the transfer of geospatial and GST learning has affected linguistic performance as well, most significantly in immigrant and second language speakers (Goldstein 2010; Goldstein and Alibrandi 2013). This transfer is only common sense, since the brain instantiates experience and information holistically, and does not segment, but connects the various phenomena to be applied in solving any number of other problems or experiences. Alibrandi and colleagues articulated a comparison of spatial and linguistic cognition, and conceptualized GIS practice as interdisciplinary educational technology (Alibrandi 2003; Thompson et al. 2000; Alibrandi et al. 2002). Some years later, the National Research Council in its report, *Learning to Think Spatially: GIS As a Support System in the K-12 Curriculum* (2006) concurred on the interdisciplinary nature and utility of GIS.

5.4 GST in Middle and Intermediate School Practice

There has been no uniform approach to integrating GST into the middle school curricula. Given the changes since early efforts and leaders, the political landscape of education has changed. The early pre-internet days of computer-based instruction depended upon the one-computer classroom, when Tom Snyder's productions, and the *Oregon Trail* software were first introduced. Many current teachers fondly recall these programs from their middle school years. The establishment and goal of KIDLINK in 1990 in Norway as part of the Children's Cultural week was to connect children between 10 and 15 years of age from all parts of the world (Wheeler 1992). When TERC and National Geographic's Kids Network became the first web-based participatory research project, many schools still hadn't joined the web as communities grappled with how computers and the internet would become part of the classroom. Later in the 1990s, ArcGIS 3.3 became available to schools, and while it was mainly private or magnet high schools that took up the technology, a few magnet middle schools did as well. Canadian schools and provincial educational organizations were quick to incorporate GST into their already robust geography and geomatics curricula (Alibrandi 2003).

5.4.1 GST in the Middle School Curriculum

Since the "middle school" concept is centered on curriculum integration (Beane 1995), the implementation and integration of GST has been varied across disciplines. Integrating GST in middle school curricula has an established history, with projects reported as early as 1999 (Thompson and Hagevik 1999; Thompson et al. 2000; Radinsky 2008; Oigara 2011). The curricular areas range from Instructional Technology to Science to Social Studies and Geography, but through curricular integration, English/Language Arts, Math (Coulter and Kerski 2005) and Economics (Radinsky 2008) have also housed substantive middle school GST. GIS was applied to oral African American history, biological assays, federal statistics, watershed studies, and GPS and geocaching to cemetery studies in American History (Oigara 2011).

5.4.2 Remote Sensing to Digital Globes

Many US science teachers had participated in NASA teacher workshops promoting RS technologies, but few were picked up at the middle school level. But these early experiences led some teachers to embrace GST before the acronym was coined. While specific middle school applications of remote sensing (RS) appear to be lacking, European use of the HERODOT imagery has aided comparative climate change studies, exemplified by the Netherlands' *Schools on Ice* project (Jekel et al. 2012) in comparative studies of climate change on glaciers. These have been more recently followed up with an increasing number of studies combining such GST as AEJEE, ArcGIS and Google Earth by middle school students for Land Use Change studies. Google Earth's *juicy geography* (http://www.juicygeography.co.uk/googleearth.htm) and Google *Lit Trips* for 6th–8th graders integrates Google Earth and young adult literature (http://www.googlelittrips.com/GoogleLit/6-8/6-8.html). There appear to be no formal studies on Google Earth use as yet.

5.4.3 GIS, GPS and Geocaching

Koolvoord and his associates in the STEM National Center for Rural Science, Technology, Engineering and Mathematics (2012), in their outreach summer workshops listed over a dozen topics for geospatial study, and had developed units for each under the general headings of Disasters, Economics, Environment and Global Warming, and Plants and animals (http://www.isat.jmu.edu/stem/curriculum.html). These activities were developed in Esri's initial online GIS application, ArcExplorer Java Edition for Education (AEJEE). In Fig. 5.1, middle school students used GPS and field log techniques to learn history in a cemetery that spanned five centuries, collecting data on various styles of cemetery markers, individuals, and historic periods. Use of GST serves multidisciplinary purposes in middle school curricula.

In their 2012 meta-analysis, Baker, Kerski, Huynh, Vehrig and Bednarz report that both early and subsequent teaching and research on middle school GST integration by Alibrandi et al. 1999, 2000; Malone 2000; Baker 2002; Coulter and Kerski 2005; Kolvoord 2008; Bednarz and Bednarz 2008; Aladag 2007, 2010; Uglurlu 2013) consistently demonstrated the active engagement of students of all ages in the GST process. Virtually all studies of middle school students have found improved achievement, attitudes and self-efficacy toward learning both

Fig. 5.1 Middle school students use GPS, geocaching and field log techniques in a cemetery spanning five centuries. Alibrandi photo, 2011

science and social studies through inquiry with GSTs (Baker 2002; Goldstein 2010; Aladag 2007; White 2006) as cited in Baker et al. (2012) review.

Baker et al. also found measurable outcomes in spatial thinking, systemic thinking competence, and scientific and geographic inquiry skills in studies of middle school students by Bednarz and Bednarz 2008; Wigglesworth 2003; Zangerl 2007; Baker 2002; and others. The Baker et al. review, also articulated measurable results on specific GST used in each of the foregoing studies. As middle level students integrated GIS, GPS and geocaching, all consummately aligned with their adolescent development, their engagement and increased competence was essentially inevitable.

5.5 GST in Practice: Framing Research in Education

In their 2012 Call for an Agenda and Center for GIS Education Research, Baker, Kerski, Huynh, Vehrig and Bednarz posed several challenges for the GST research community. At the same time, they presented a comprehensive matrix of research that has been published in the field of GSTs in education. Among their selected research questions were:

- What impact does GIS have on the ability to think critically, not only in geography, but in other disciplines?
- Does GIS in education impact skills that are measurable in any type of assessment, standardised or otherwise?

- How effective is GIS for males versus females, for special needs students, for students where the software or instruction is not in their native language, for at-risk students?

5.5.1 Case Study: Research Findings on Middle School GST Yield Standardized Test Results

Our recent (2013) reporting of Goldsten's (2010) research relates to each of the questions above. In her study of middle school students enrolled in a partial-year GIS elective class to which 90 % of the students had been randomly assigned, results on the state's standardized reading test revealed surprising results for the GIS students. All students who had taken the GS elective had elevated reading test scores on the Florida Comprehensive Achievement Test (FCAT). The reading test did not test for geographic content, but for reading comprehension. That the GIS elective students outperformed their non-GIS colleagues demonstrated significant transfer of knowledge from the sophisticated GIS technology to their application of that knowledge to general reading comprehension.

If reading is central to learning, and if GIS integration positively impacts reading scores, this learning may be transferred to other disciplines as well. Students may construct new knowledge in other subjects by building upon the gains in reading related to the GIS instruction. This outcome may also be a residual effect of the students actually being engaged in the learning and motivated to comprehend and apply new knowledge. As their culminating assessment, students in the GIS elective class employed problem- and place-based learning on local GIS projects. For these, students generated and applied new environmental (GPed turtle nesting sites) and local census data. All of the GIS students' grades in science and social studies classes improved during the GIS semester (Goldstein 2010). Findings from the standardized test are presented in Table 5.1.

These findings had even greater significance among the highly diverse district's Non-Native English speaking GIS student scores that were significantly higher than

Table 5.1 Effect of groups on moderation of primary language on FCAT reading scores by primary language

Source	Group	*M*	*SD*	*N*	*t*	*df*	*p*	η^2
Non-native english speaking students								
FCAT reading	GIS	340.27	41.82	44	12.98	380	0.000*	0.307
	Non-GIS	246.46	45.49	338				
Native english speaking students								
FCAT reading	GIS	344.38	47.33	133	16.2	917	0.000*	0.222
	Non GIS	260.09	56.77	786				

Source: From Goldstein (2010)
$*p < 0.025$

those of their non-GIS cohort. In fact, they were within four points of the first language English speaking GIS students. Subsequent observations of the above students, as well as and research in second language learning yielded supporting findings by Pan and Pan (2009) on the effects of just "pictures" in (English Language Learners (ELLs). The transfer phenomenon is more broadly supported by Brinton's highly cited works contending that when complex information is delivered through real life context, ELL students make greater connections with language and prior knowledge. When content-based information (CBI) is reiterated strategically; i.e., in the constant *application* of the new language at the right times for its utility, it compels students to learn from their passion and interest in *applying* the new language along with their spatial thinking skills (Brinton 2003).

In GIS class, students do not depend solely on the teacher, but interact with their peers to construct multiple types of new knowledge *applied* to spatial tasks (Goldstein and Alibrandi 2013). The necessity of collaboration and use of the new terms greatly increased the linguistic capacity of all students engaged in a multidisciplinary problem, project and place-based learning application such as that of the GST classroom.

Goldstein's (2010) findings amplify those from a multistate study. Hinde et al. (2011) found improved ELLs' performance on literacy tests:

> Multistate research revealed significant increases in reading comprehension in most elementary grades from the teaching of *GeoLiteracy* lessons. These findings are consistent with a body of evidence revealing that integrating the curriculum increases student achievement in the tested skill areas of math and reading .... (Hinde et al. 2011)

Too often, the results of middle and intermediate school GST integration have gone unmeasured because both science and geography (which, as in the US) go untested. It would therefore be critical to encourage results-based studies in nations where not only geography and GSTs are present and tested on standardized tests, but in developing, as Baker et al. (2012) have articulated, ways of measuring the transfer benefits across the curriculum as well as developing measures of problem-solving.

5.6 Discussion: Implications and Conclusions: GST as Ontogenic Twenty-First Century Skills

As this volume comes to print, GIS software developer Esri US has released its online version, *ArcGISOline* (AGO) and has made available licenses to all US schools though President Obama's ConnectEd initiative, finally incorporating the power of the internet with its analytic capability. Europe's Schoolnet will initiate dissemination, and Esri Canada is establishing Centers for its K-12 support.

In our experience that spans several US school districts, states and Canadian examples, collaboration has been central to the success and sustainability of GST in schools. Partnership is critical to support teachers and students in understanding and applying GSTs to school and community as well as curricular goals. The US state of

Delaware's Delaware Geoeduction partnership, composed of the Delaware Center for Educational Technology, Delaware Geographic Alliance, the Delaware GIS Day Committee, Delaware Technical and Community College, Delaware Geographic Data Committee, the Delaware Department of Education, and Esri GIS for Schools has combined efforts with statewide funding partners to promote GIS and geospatial education in a collaborative model that would benefit further study by any political unit. Several of the states teachers have received awards for their work to promote Geoeducation (http://mygeoworld.org/our-partners). This level of collaboration, similar to that of many other GST successful institutional collaborations, is critical to the sustainability of GST in schools.

Since 2009, the re-institutionalization of Esri's early (1998–1999) invited summer teacher training workshops have been expanded to larger groups of participants (T3G) and repeated annually for outreach and training to educators who can reach a variety of potential GST learners in the US and Canada.

We must consider the lifespans of today's Middle and Intermediate school students. Their lives began in the twenty-first century. It is time to let go of the "ownership" of GST teaching and learning by disciplines. Rather, GST must now be seen as central to the networked, connected and interdisciplinary convergences of the integrated curriculum. We recommend the following:

1. The GST community must articulate GSTs' centrality to twenty-first century solutions;
2. Promote GST as the perfect *developmentally appropriate integrative approach* for adolescent learners;
3. Facilitate a societal construct for imagining the landscapes of learning that have yet to unfold for this century;
4. Collaboration between classroom and community (aka "real world") be seen as the new landscape of learning

The significance of GSTs to make not just visible, but *possible* the new sciences (climatic, paleoclimatic, biological, ecological, environmental, space, etc) of the twenty-first century, is a break with past discipline-based constraints. The constraints must become integrated, as in the middle school curriculum, as a failure to make this leap is essentially analogous to avoiding the next enlightenment. The resistance GST educators have experienced is the prevailing fear of moving beyond the bounds of the disciplines. But flexibility and the integrated middle school curriculum should be precisely the developmental and curricular target for modeling the integration, connection and applications across the whole curriculum. This is nothing short of facilitating the sea change already in motion that will become the next wave of learning and teaching as a naturally integrated, interdisciplinary landscape in which humans have evolved. As GST can combine such diverse data as biology, paleoclimatology, genetic research, economics, geology, archeology and new energy applications, so must the ontogenesis be modeled in schools. GST has become a way to imagine the future, and that lies for today's middle school students, across the duration of the twenty-first century.

References

Aladag, E. (2007). *The effects of using GIS on students' academic achievement and their motivations towards the lesson in the 7th grade social science course in primary education.* Unpublished PhD dissertation, Gazi University, Ankara, TK.

Aladag, E. (2010). The effects of GIS on students' academic achievement and motivation in seventh grade social studies lessons in Turkey. *International Research in Geographical and Environmental Education, 19*(1), 11–23.

Alibrandi, M. (2003). *GIS in the classroom: Using geographic information systems in social studies and environmental science*. Portsmouth: Heinemann.

Alibrandi, M., Thompson, A., & Hagevik, R. (1999, July). *The ligon history project: Exploring the past to influence the future*. Paper Presented at the Education User Conference (EdUC). ESRI, San Diego, CA. http://ced.ncsu.edu/ligon/about/history/esri/P7311.htm

Alibrandi, M., Beal, C., Wilson, A., & Thompson, A. (2000). Reconstructing a school's history using oral histories and GIS mapping. *Social Education, 64*(3), 134–140.

Alibrandi, M., White, S., Hagevik, R., & Duke, B. (2002). *GIS in teacher education: Notes from the field II*. Paper presented at the paper presented at the Education User Conference (EdUC). ESRI, San Diego, CA.

Ash-Duke, B. (2005). Building a successful GIS program in a middle school. In C. Crawford et al. (Eds.), *Proceedings of society for information technology & teacher education international conference 2005* (pp. 3547–3552). Chesapeake: AACE. http://www.barbareeduke.com/downloads/BuildGISProgMS.pdf

Baker, T. R. (2002). *The effects of geographic information system (GIS) technologies on students' attitudes, self-efficacy, and achievement in the middle school classroom.* Unpublished PhD dissertation, University of Kansas.

Baker, T., Kerski, J., Huynh, N., Vehrig, K., & Bednarz, S. (2012, Winter). *Review of International Geographical Education Online*, *2*(3). http://www.rigeo.org/vol2no3/RIGEO-V2-N3-1.pdf

Beane, J. A. (1995, April). Curriculum integration and the disciplines of knowledge. *The Phi Delta Kappan, 76*(8), 616–622. http://www.jstor.org/stable/20405413

Bednarz, S., & Bednarz, R. (2008). Spatial thinking: The key to success in using geospatial technologies in the social studies classroom. In A. Milson & M. Alibrandi (Eds.), *Digital geography: Geo-spatial technologies in the social studies classroom.* Charlotte: Information Age.

Brinton, D. (2003). Content – Based instruction. In Nunan (Ed.), *Practical English language teaching* (pp. 199–224). New York: McGraw-Hill.

Callahan, C. M., & Hertberg-Davis, H. L. (2013). *Fundamentals of gifted education: Considering multiple perspectives.* London: Routledge.

Coulter, B., & Kerski, J. (2005). Using GIS to transform the mathematical landscape. In *Technology-supported mathematical learning environments: 67th yearbook of the national council of teachers of mathematics.* Reston: National Council for Teachers of Mathematics (NCTM).

Downs, R. M., & Stea, D. (1977). *Maps in minds: Reflections on cognitive mapping*. New York: Harper & Row.

Gardner, H. (1985). *Frames of mind: The theory of multiple intelligences.* New York: Basic Books.

Gardner, H. (2011). *The unschooled mind: How children think and how schools should teach.* Basic Books.

Giedd, J. (2004). Structural magnetic resonance imaging of the adolescent brain. *Annals of the New York Academy of Science, 1021*, 77–85. NY Academy of Science.

Goldstein, D. (2010). *Integration of geospatial technologies into K-12 curriculum: An investigation of teacher and student perceptions and student academic achievement*. Unpublished PhD dissertation, Florida Atlantic University, 2010 (UMI 3406541).

Goldstein, D., & Alibrandi, M. (2013). Integrating GIS in the middle school curriculum: Impacts on diverse students' standardized test scores. *Journal of Geography, 112*(2), 68–74.

Gould, P., & White, R. (1974). *Mental maps*. Harmondsworth: Penguin.

Hinde, E. R. (2005). Revisiting curriculum integration: A fresh look at an old idea. *The Social Studies, 96*(3), 105–111.

Hinde, E. R., Popp, S., Dorn, R., Ekiss, G., Mater, M., Smith, C., & Libbee, M. (2007). The integration of literacy and geography: The Arizona GeoLiteracy program's effect on reading comprehension. *Theory & Research in Social Education, 35*(3), 343–365.

Hinde, E., Osborn Popp, S., Jimenez-Silva, M., & Dorn, R. (2011). Linking geography to reading and English language learners' achievement in US elementary and middle school classrooms. *International Research in Geographical and Environmental Education, 20*(1), 47–63.

Jekel, T., Koller, A., & Strobl, J. (2012). Research institutions linking to schools: Collaboration for geoinformation in secondary education in Austria. In A. Milson, A. Demirci, & J. Kerski (Eds.), *International perspectives on teaching and learning with GIS in secondary schools*. New York: Springer.

Kolvoord, R. (2008, March/April). Geospatial technologies: Real projects in real classrooms. *Knowledge Quest Maps the Why of Where, 36*(4). http://www-tc.pbs.org/teachers/librarymedia/aasl/kolvoord.pdf

Kolvoord, R. (2012). *Curriculum page: STEM national center for rural science, technology, engineering & mathematics education outreach geospatial technology*. http://www.isat.jmu.edu/stem/curriculum.html

Malone, L. (2000). *Barrington middle school*. http://www.esri.com/news/arcnews/fall00articles/comatlas.html

National Research Council. (2006). *Learning to think spatially: GIS as a support system in the K-12 curriculum*. Washington, DC: National Academies Press.

Oigara, J. (2011). Using GPS technology and geocaching as a method to teach social studies curriculum. In M. Koehler & P. Mishra (Eds.), *Proceedings of society for information technology & teacher education international conference 2011* (pp. 3095–3096). Chesapeake: AACE.

Pan, Y., & Pan, Y. (2009). The effects of pictures on the reading comprehension of low-proficiency Taiwanese English foreign language college students: An action research study. *VNU Journal of Science, Foreign Languages, 25*(4), 186–198. http://js.vnu.edu.vn/nn 3 09/b.8.pdf

Radinsky, J. (2008). Digital geography: Geo-spatial technologies in the social studies class room. Charlotte: Information Age. http://www.gisforhistory.org/info/abouttheproject.php

Renzulli, J., & Reis, S. (2013). Chapter 21: The schoolwide enrichment model: A focus on student creative productivity, strengths, and interests. In C. Callahan & H. Herthberg-Davis (Eds.), *Fundamentals of gifted education: Considering multiple persepectives*. New York: Routledge.

Thompson, A., & Hagevik, R. (1999). *Satellites, computers, and mapping: Syllabus and student projects*. http://www.ncsu.edu/midlink/gis/courseoutline.htm

Thompson, A., Alibrandi, M., & Hagevik, R. (2000). Remaking history with interdisciplinary GIS. In G. Ludwig & R. Audet (Eds.), *GIS in schools*. Redlands: Environmental Science Research Institute.

Uglurlu, N. B. (2013). *Middle school students' opinions towards geographic information systems in Turkey*. Adiyaman Üniversitesi Sosyal Bilimler Enstitusu Dergisi. ISSN: 1308–9196 Yıl : 6 Sayı : 14 Sosyal Bilgiler Ögretimi Özel Sayisi. 2013. http://www.adyusbd.com/Makaleler/745606151_BALO%C4%9ELU-%20N%C4%B0HAL.pdf

UN World Youth Report. (2013). http://undesadspd.org/WorldYouthReport/2013.aspx

Wheeler, D. (1992, April 22–24). *KIDLINK in war and peace*. Paper presented at the annual conference of the American Educational Research Association, San Francisco, CA. http://files.eric.ed.gov/fulltext/ED350985.pdf

White, S. H. (2006). *Geographic information systems (GIS) and instructional technology (IT) diffusion: K-12 student and educator conceptualizations*. Unpublished dissertation. http://www.lib.ncsu.edu/theses/available/etd-03142003-001503/unrestricted/etd.pdf

Wigglesworth, J. (2003). What is the best route? Route-finding strategies of middle-school students using GIS. *Journal of Geography, 102*(6), 282.

Zangerl, S. (2007). Navigation auf virtuellen Globen. In T. Jekel, A. Koller, & J. Strobl (Eds.), *Lernen mit Geoinformation II* (pp. 106–115). Heidelberg: Wichmann.

Chapter 6
Geospatial Education in High Schools: Curriculums, Methodologies, and Practices

Che-Ming Chen and Yao-Hui Wang

Abstract Nowadays high school education throughout the world is expected to prepare the young people for understanding the twenty-first century globalization and adapting themselves in this fast-changing world. The ability of applying spatial thinking to real-world issues at a range of scales is essential for a responsible modern citizen and for an effective knowledge worker as well. Educators and students who can use geospatial technologies are more efficient spatial thinkers. That's why the geospatial education in high schools receives increasing attention worldwide.

The adoption of geospatial technologies in high school education varies among and within countries in term of the status of geography course, information infrastructure, and teacher preparation. There is no general model toward the success in geospatial education. This paper focuses on finding the exemplars of curriculums, methodologies, and practices particularly suited to geospatial education in high school. Besides, GIS, GPS, and RS are the instructional tools traditionally used in geospatial education. The emergences of new technologies such as social media, interactive web-mapping platforms, and smartphones allow individuals or social groups to collect, georeference, and share geographic information. The applications of these technologies and their spatial data in high school geospatial education are also addressed.

Keywords High school education • Geospatial technologies (GSTs) • Geospatial education

6.1 Introduction

When high schools incorporate geospatial technologies (GSTs) into their learning activities, they may get students involved in collaborative learning and expand their learning opportunities beyond the classroom. Sarah Chase-Walsh and Noah

C.-M. Chen (✉) • Y.-H. Wang
Department of Geography, National Taiwan Normal University, Taipei, Taiwan
e-mail: jeremy@ntnu.edu.tw; linkgis@gmail.com

O. Muñiz Solari et al. (eds.), *Geospatial Technologies and Geography Education in a Changing World*, Advances in Geographical and Environmental Sciences,
DOI 10.1007/978-4-431-55519-3_6

Fig. 6.1 Sarah and Noah demonstrates how their mobile app works (ESRI Video 2012)

Pilchen, two 12-grade students of Washington-Lee High School in Virginia, shared their Geographic Information Systems (GIS) project work in the 2012 ESRI International User Conference Plenary. A local storm water administration had to apply "No Dumping" labels to every storm sewer in the city. They used to keep track of the sewer's label status using a highlighter and a wall-sized paper map. Sarah and Noah not only replaced that wall map to a digital map for them using ArcMap, but also created an online map and a mobile app (Fig. 6.1). Now their specialist can simply bring an iPad, and use a pull-down menu in the app to choose one of the three labels, "Unmarked", "Some Marked", or "Fully Marked". The label status of sewers in the online map will simultaneously be updated when the editing is made in the field (ESRI Video 2012).

6.2 What Key Skills Are We Expecting Our High School Students to Have?

Sarah and Noah represent the next generation of GSTs professionals. They see a real-world problem and use GSTs to solve it. Nowadays high school education throughout the world is expected to prepare young people for understanding the twenty-first century globalization and adapting themselves in this fast-changing world. The ability of applying spatial thinking to real-world issues at a range of scales is essential for a responsible modern citizen and for an effective knowledge worker as well. Educators and students who can use GSTs are more efficient spatial thinkers (Bednarz and Bednarz 2008).

Our twenty-first century high school students are digital natives. GSTs are commonly used in their daily live. For example, mobile phones allow them to share location information and geotagged photos. Many spatial decisions are made by the use of Google Maps or Google Earth such as planning a tour, finding the shortest path to a restaurant, or avoiding traffic jams. GSTs also promote their learning in school. Such tools can engage them into spatial thinking and bring advantages in learning subjects such as geography, history, science, and mathematics (Sanchez 2009). In addition to the practical value in day-to-day life and school, GSTs have been recognized as high growth industry and the related practitioners are increasingly needed in the job markets (Nielsen et al. 2011). In the era of globalization, educators worldwide are seeking the solutions for the education of modern citizens. Spatial thinking is considered as the fundamental skill for understanding globalization. However, spatial thinking skills are less easy to be developed by the traditional pedagogies.

The contemporary educational paradigm is the learning theory of constructivism, which claims that knowledge is constructed by learners based on one's own experiences in authentic context rather than transferred from teacher to student. Spatial technologies can empower students to engage in inquiry-based or problem-based learning about local to global issues, and match the constructivist's teaching style (Demirci et al. 2013; Milson and Kerski 2012). The benefits of the use of spatial technologies in high school education might be considered as the *3Cs* – to their communities, to citizenship education, and to careers (Kerski et al. 2013).

The effectiveness of GSTs on high school students' learning have been validated by many empirical researches in the last decade (Patterson et al. 2003; West 2003; Kerski 2003; Demirci et al. 2011, 2013). However, the adoption of GSTs in high school education varies among and within countries in term of the status of geography course, information infrastructure, teacher preparation, and other factors. There is no general model toward the success in geospatial education. This chapter will focus on providing the exemplars of curriculums, methodologies, and practices particularly suited to geospatial education in high schools. Besides, GIS, Global Positioning System (GPS), and Remote Sensing (RS) are the instructional tools traditionally used in geospatial education. The emergences of new technologies such as interactive web-mapping platforms and mobile devices would allow individuals or social groups to receive, collect, georeference, and share geographic information. The inclusion of these technologies and their spatial data in high school geospatial education are also addressed in the examples of practices.

6.3 GST in High School Curriculums

GST can be integrated into high school education in many subjects such as geography, history, economics, and mathematics, etc. However, geography is the most relevant subject. GIS is included into national curriculums of geography in many countries as a compulsory part, an optional unit, or a selective course. The depth of GIS in the geography standards can be classified as two levels: GIS awareness and GIS application. The GIS awareness means the standards explicitly mention GIS, but introduce only GIS concepts and functions. The standards of GIS application further expect students to use GIS for higher-order thinking such as problem solving or decision making (Milson and Roberts 2008).

For example, the national curriculum of high school geography in South Africa provides a typical case of GIS awareness. Its standard provides the introduction of GIS functions in great detail. The principle of GIS is introduced at 10th grade. GIS functionalities such as data acquisition including remote sensing, data preprocessing and post processing are covered at 11th grade. At the final year of high school, advanced GIS functionalities are taught including data management, data manipulation, data analysis, spatial analysis, spatial statistics, map production, and applications (Eksteen et al. 2012). The national geography standard of Taiwan provides an example of initiation on GIS awareness and then moves to GIS application. Geography is a compulsory subject at Grade 10–11 in Taiwan. The GIS class at Grade 10 follows up the practical use of GIS in students' everyday lives with the basics of GIS. The guideline suggests students take 1–2 lab hours to have hands-on experience of GIS software. Students are expected to either use professional desktop GIS or free Web-based GIS like Google Earth to query about geographic issues around the local community and to demonstrate the results. GPS is also recommended to be used in collecting field data. The various applications of GIS are further introduced in geography at Grade 12 which demonstrate how GIS can be applied to monitor mudslides, diseases, floods, and urban planning (Lay et al. 2013; Wang and Chen 2013).

Rather than placing emphasis on understanding the concepts and power of GIS, several national or state curriculum standards regard GIS as a tool for higher-order thinking, problem solving, or decision-making. For example, GIS is included in the national geography standard of upper secondary school in Norway. The geography course with 2 hours per week is compulsory for all 1st grade students. Students should know how to use digital tools like GIS to collect, measure, manipulate, and visualize geoinformation (Rød et al. 2012). In the United States, The Texas high school geography standards expect students to collect a variety of sources including geographic information to answer geographic questions, infer geographic relationships, analyze geographic change, and solve geographic problems. The creators of the New York Standards expect students to analyze geographical information in a spatial database or GIS, represent geographical inquiry in maps, develop hypotheses, test hypotheses and formulate conclusions (Milson and Roberts 2008). To sum up, the awareness of GIS is better than the lack of GIS in geography standards of high schools. However, only engaging students into real-world problems can help them to experience the power of GST and make them a full-fledged modern citizen with higher-order thinking skills.

6.4 Teaching Methods and Effective Practices

Teaching methods very widely and there are numerous studies in the literature dealing with the classification of teaching methods. Most of the taxonomies tend to classify the methods into several categories. For example, Clark (2008) suggests that teaching methods could be classified into four categories, including receptive instruction, directive instruction, guided discovery and exploratory instruction. Fenstermacher and Soltis (2004) classify different methods into three categories by teaching style, consisting of teacher as executive, facilitator or liberationist. However, Joyce et al. (2009) are among the leaders in the classification of teaching methods. They define four major families of teaching methods, comprising behavioral systems family, information-processing family, personal family and social family. Their model has been well-accepted by educational researchers.

In order to describe the practices of geospatial education in high schools comprehensively, the model contributed by Joyce, Weil, and Calhoun is used to provide the framework for the following discussion. Most of the practices fall on behavioral system family, information-processing family, and social family. It doesn't mean one family is superior to others, because a teacher may adopt more than one family to meet different learning goals. Besides, the effectiveness of a family to some extent is dependent on cultural contexts. The examples of practices will be introduced following each family.

6.4.1 Behavioral System Family

The behavioral system family attempts to build efficient environments for shaping learner's behaviors by manipulating reinforcement. The examples of the methods in the family are direct instruction and mastery learning. These methods tend to be teacher-centered. They focus on observable skills and behaviors. It has been proven that these methods could be more effective than models in other families for increasing the scores on standardized tests of basic skills (Huitt 2003). They are especially effective when there are a great number of students in the classroom or the syllabus hour is very limited. Many researchers also found that teacher-centered instructions seemed favorable for teachers in many Asian countries, because student-centered instructions, such as problem-based learning and cooperative learning are more difficult to be implemented (Yap et al. 2008; Lam et al. 2009; Yuda 2009).

Teacher-centered instruction is frequently applied to introduce the concepts of GIS and to learn how to operate GIS software. For example, from a national-wide survey of high school geography teachers in Taiwan, Wang and Chen (2013) found that 93 % of the teachers choose didactic and catechetical instruction as their primary teaching methods when they "teach about" GIS. In the introductory GIS lessons, the textbook publisher's CDs are the major supplementary resource. The

CDs contains animations and slides which are efficient for demonstrating the concepts and basic functions of GIS and make them popular for teachers. The behavioral system family seems a logical choice for the geospatial education in the entry level especially when resources are short and class size is big. However, students will need more hands-on exercises and inquiry-based learning when higher-order thinking skills are expected (Dong and Lin 2012).

6.4.2 Information-Processing Family

The models of the information-processing family aim to help learners to acquire data, sense problems, generate solutions, develop concepts, and employ verbal or non-verbal symbols for communication. Inquiry-based learning (IBL) and problem-based learning (PBL) are two strategies frequently cited in research. Both in IBL and PBL, group work is a common feature. Students gather together to reflect critically, and they collaborate to develop questions, investigate information, construct knowledge, and share understandings of information. Although the two methods are very similar, there are some slight differences. For example, the knowledge to be developed is often acquired before the investigation takes place in IBL. Therefore, in-class learning is prior to undertaking fieldwork. In contrast, learning activities in PBL are "problem first". The problem is usually set by the teacher in the first place for stimulating learner's motivation (Spencer and Jordan 1999).

Favier and van der Schee (2012) provide a success story for applying IBL with GST to a student research project. After the introduction of economic-geographic concepts like "range" and "market area", students were required to choose a particular service in town for case study. They had a short GIS training session and learned how to draw a map representing the market area of a service. Students chose four local gyms as the targets for investigation. They were asked to formulate hypotheses about the order of their market area size. Then they went to the gyms and interviewed 20 customers at each location. Students later marked the customers' home location in the map and evaluated the agreement between the reality and their assumption. They presented their research results with maps to their teacher and classmates. The teacher followed up their presentation with discussion and summarized what they learned.

Hsu and Chen (2010) provided another example of IBL practice by integrating virtual reality and mobile learning for implementation. Fourteen students as five teams were asked to evaluate three proposals for improving a polluted irrigation ditch in the local area. A virtual field trip (VFT) website was established allowing students to learn the background information of this irrigation ditch, to understand the basic concepts of sustainable development, and to virtually explore the surroundings of each fieldwork site. Then the students conducted a self-guided fieldtrip with a PDA which provided the functions of field navigation and learning support. Students were navigated by the GPS in the PDA to investigate 5 fieldwork sites. The

PDA would automatically prompt students to proceed to fieldwork assignments when they are within 15 m of the 5 fieldwork sites. For example, the PDA asked students to interview two local residents for sharing their visions about the irrigation ditch, and to save the results as text message in the system. After the completion of fieldworks, students downloaded the field data from PDA, evaluated the proposals based on their field investigation, picked one out of three proposals, and validated their decision. The feedback from teachers and students shown that teachers highly regarded the VFT and PDA as useful tools for facilitating fieldwork and students felt they are true "scientists" because they could complete the research with only little help from teachers.

Liu et al. (2010) developed PBL learning activities with and without GIS support to validate the assumption that learning with GIS can result in higher-order learning outcomes. In the first place, an ill-structured problem was provided to students regarding to the significant growth of migrants and its implications with the aging society in Singapore's population. Students had to play as the Head of the Research Team in the Aging Population Committee and answer the questions such as "Any specific town, constituency, or zone that has a high number of elderly?", "Where would you think the Government might need to provide more aged care facilities and services?", "At which Mass Rapid Transit stations would you expect the greatest numbers of escalator accidents involving the elderly?", and "What facilities need to be provided to avert this problem?". The students in the experimental group used ArcGIS and geographic data including provincial boundaries, census data, socioeconomic data, and satellite imagery to find the answers of the above questions. The results of this quasi-experimental research proved the experimental group using GIS outperformed the control group without using GIS in the tests of higher-order thinking skills.

6.4.3 Social Family

The teaching models of the social family encourage students to build learning community so that they can work together and learn from each other. Therefore, one general educational outcome here is the development of solid citizenship (Joyce et al. 2009). Geographic problems in the real world are often complex, and solving the problems usually require Interdisciplinary team work. The social family could improve student's learning motivation and interpersonal skills which are important in the workplace. Role playing and collaborative learning are two common pedagogical models in this family.

Sanchez et al. (2010) designed a role-playing game about sustainable development and used GST to engage students in complex situations in the real world. The game starts from the project call by the city mayor (teacher). Six companies (6 pairs of students) have to design a project for implementing new energies in the city of Sète, south of France. Each company specialized in different sorts of green energies including heat pumps, windmills, ocean wave energy, photovoltaics and methanization. The committee tender (pair of students) is responsible for the process rules and to consult a local association of citizens (another pair of students) for the best choice among six projects. Students playing the six companies use Google Earth to provide 3D model with a site view for their project and to assess the impact on the local environment. The committee tender and the association of citizens have to go to the field with a GPS embedded Pocket PC to collect data for validating the projects. The interview videos of local residents are provided to them in different places in the city via MITAR, augmented reality software. The students use posters to present their final project in the school library. Finally all students play as local residents and vote for the best project via a website. In such a game-based learning design, GSTs permit teachers to design complex situations in real world and engage students to a problem requiring multidisciplinary knowledge and communication skill to solve.

The Scientific Research Class of Red Bank High School in Chattanooga Tennessee, USA presents a success story about the collaborative project among high school students, OpenStreetMap volunteers, and GIS Corps. Padang is one of Indonesia's most vulnerable cities. An earthquake in 2009 claimed over 1,100 lives in that area and over 800,000 people are at risk from earthquake and tsunami activities in the future. The students in the high school worked as a team to digitize all of the visible buildings and roads in the satellite imagery. The online map in great detail allows the local government and others to estimate the number of people and buildings that will be impacted by natural disasters. The interactive web-mapping platforms like OpenStreetMap could connect students to the rest of world and expand their learning beyond school (Hale 2012).

6.5 Conclusions

In this article, we illustrated a comprehensive curriculum standard of GSTs in high school geography which should include the basic concepts about tools and apply them to solve problems in the real world so that the spatial thinking skills could be developed. There are various methods for the implementation of GSTs in the lessons. The behavioral system family is effective for understanding the basics of spatial technologies and for hands-on training of software in a large group of students. However, if the teachers relying solely on behavioral system family methods want to make the learning more fun and incorporate more higher-order thinking skills, a blended approach among different families could be a suitable option. The information-processing family could move the teaching beyond the lower levels of Bloom's taxonomy by engaging students into scientific enquiry or solving real-world problems with GSTs. But we could not ignore the disadvantages associated with information-processing family methods. They include the increased time commitment of instruction and problems associated with group work. In order to maintain students' concentration and motivation while studying, it is recommended that the selected themes should be engaging and personally meaningful to students. The social family tries to replicate the complex situations in the real world and simulate how people interact with co-workers. However, solving geographic problems in the real world usually requires interdisciplinary team work. It is essential for geography teachers to invite teachers from other subjects or experts outside schools to involve students' learning process. In such ways, the education of spatial technologies in high schools could better match the needs of context-specific knowledge and skills in a workplace. Hopefully, the examples of practices following up with each method will extend their spheres of influence and inspire more innovative lesson plans to be created.

References

Bednarz, S. W., & Bednarz, R. S. (2008). Spatial thinking: The key to success in using geospatial technologies in the social studies classroom. In A. J. Milson & M. Alibrandi (Eds.), *Digital geography: Geospatial technologies in the social studies classroom* (pp. 249–270). Charlotte: Information Age Publishing.

Clark, R. C. (2008). *Building expertise: Cognitive methods for training and performance improvement*. San Francisco: Pfeiffer.

Demirci, A., Karaburun, A., Ünlü, M., & Ozey, R. (2011). Using GIS-based projects in learning: Students help disabled pedestrians in their school district. *European Journal of Geography, 22*, 48–61.

Demirci, A., Karaburun, A., & Ünlü, M. (2013). Implementation and effectiveness of GIS-based projects in secondary schools. *Journal of Geography, 112*(5), 214–228.

Dong, P., & Lin, P. (2012). China: Teacher preparation for GIS in the national geography curriculum. In A. J. Milson, A. Demirci, & J. J. Kerski (Eds.), *International perspectives on teaching and learning with GIS in secondary schools* (pp. 59–64). Dordrecht: Springer.

Eksteen, S., Pretorius, E., & Breetzke, G. (2012). South Africa: Teaching geography with GIS across diverse technological contexts. In A. J. Milson, A. Demirci, & J. J. Kerski (Eds.), *International perspectives on teaching and learning with GIS in secondary schools* (pp. 225–232). Dordrecht: Springer.

ESRI Video. (2012). *Sarah Chase-Walsh and Noah Pilchen of Washington-Lee High School share their GIS project work*. http://media.esri.com/arcstream/2012/08/1636-washington_dash_lee-high-school-uses-problem_dash_based-learning-and-arcgis_960.mp4. Accessed 25 Aug 2014.

Favier, T. T., & Van der Schee, J. A. (2012). Exploring the characteristics of an optimal design for inquiry-based geography education with Geographic Information Systems. *Computers & Education, 58*(1), 666–677.

Fenstermacher, G. D., & Soltis, J. F. (2004). *Approaches to teaching*. New York: Teachers College Press.

Hale, R. (2012). *Red Bank high school and HOT*. http://www.northrivergeographic.com/archives/red-bank-high-school-and-hot. Accessed 25 Aug 2014.

Hsu, T. Y., & Chen, C. M. (2010). A mobile learning module for high school fieldwork. *Journal of Geography, 109*(4), 141–149.

Huitt, W. (2003). Models of teaching/instruction. *Educational Psychology Interactive*. http://chiron.valdosta.edu/whuitt/col/instruct/instmdls.html. Accessed 25 Aug 2014.

Joyce, B., Weil, W., & Calhoun, E. (2009). *Models of teaching* (8th ed.). Boston: Allyn and Bacon.

Kerski, J. J. (2003). The implementation and effectiveness of geographic information systems technology and methods in secondary education. *Journal of Geography, 102*(3), 128–137.

Kerski, J. J., Demirci, A., & Milson, A. J. (2013). The global landscape of GIS in secondary education. *Journal of Geography, 112*(6), 232–247.

Lam, C. C., Lai, E., & Wong, J. (2009). Implementation of Geographic Information System (GIS) in secondary geography curriculum in Hong Kong: Current situations and future directions. *International Research in Geographical and Environmental Education, 18*(1), 57–74.

Lay, J. G., Chen, Y. W., & Chi, Y. L. (2013). GIS adoption among senior high school geography teachers in Taiwan. *Journal of Geography, 112*(3), 120–130.

Liu, Y., Bui, E. N., Chang, C. H., & Lossman, H. G. (2010). PBL-GIS in secondary geography education: Does it result in higher-order learning outcomes? *Journal of Geography, 109*(4), 150–158.

Milson, A. J., & Kerski, J. J. (2012). Around the world with geospatial technologies. *Social Education, 76*(2), 105–108.

Milson, A. J., & Roberts, J. A. (2008). The status of geospatial technologies in US high school geography standards. In A. J. Milson & M. Alibrandi (Eds.), *Digital geography: Geospatial technologies in the social studies classroom* (pp. 39–59). Charlotte: Information Age Publishing.

Nielsen, C. P., Oberle, A., & Sugumaran, R. (2011). Implementing a high school level geospatial technologies and spatial thinking course. *Journal of Geography, 110*(2), 60–69.

Patterson, M. W., Reeve, K., & Page, D. (2003). Integrating Geographic Information Systems into the secondary curricula. *Journal of Geography, 102*(6), 275–281.

Rød, J., Andersland, S., & Knudsen, A. (2012). Norway: National curriculum mandates and the promise of Web-based GIS applications. In A. J. Milson, A. Demirci, & J. J. Kerski (Eds.), *International perspectives on teaching and learning with GIS in secondary schools* (pp. 191–199). Dordrecht: Springer.

Sanchez, E. (2009). Innovative teaching/learning with geotechnologies in secondary education. In *Education and technology for a better world* (pp. 65–74). Heidelberg: Springer.

Sanchez, E., Delorme, L., Jouneau-Sion, C., & Prat, A. (2010). Designing a pretend game with geotechnologies: Toward active citizenship. In T. Jekel, A. Koller, K. Donert, & R. Vogler (Eds.), *Learning with geoinformation V* (pp. 31–40). Heidelberg: Wichman.

Spencer, J. A., & Jordan, R. K. (1999). Learner centred approaches in medical education. *British Medical Journal, 318*(7193), 1280–1283.

Wang, Y. H., & Chen, C. M. (2013). GIS education in Taiwanese senior high schools: A national survey among geography teachers. *Journal of Geography, 112*(2), 75–84.

West, B. A. (2003). Student attitudes and the impact of GIS on thinking skills and motivation. *Journal of Geography, 102*(6), 267–274.

Yap, L. Y., Ivy Tan, G. C., Zhu, X., & Wettasinghe, M. C. (2008). An assessment of the use of Geographical Information Systems (GIS) in teaching geography in Singapore schools. *Journal of Geography, 107*(2), 52–60.

Yuda, M. (2009). *Study on utilization of geographic information system in school education in Japan*. Doctoral dissertation. http://dspace.lib.kanazawa-u.ac.jp/dspace/bitstream/2297/17573/1/Thesis-YUDA-M-252.pdf. Accessed 25 Aug 2014.

Chapter 7
Applied Geospatial Technologies in Higher Education

Reed Perkins

Abstract This chapter discusses the international trends, roles, and instruction of GST in higher education. While all countries continue to emphasize teaching *about* GST, many are developing strategies to teach *with* GST. Technological advances (cloud-based GST applications, increasingly sophisticated software, etc.) are driving changes in higher education GST objectives, communication, and pedagogy. Countries are developing institutional and faculty peer networks to strengthen the depth and consistency of their GST curricula, as well as keep pace with rapid changes in GST itself. Best pedagogical practices for GST instruction are being developed, including high-impact techniques such as collaborative projects, undergraduate research and inquiry, global learning, and community-based learning. Challenges facing GST in higher education include limited availability of GST courses of study and student enrollment in GST programs. The rapid rate of change in GST technology is a significant barrier to faculty preparation, especially regarding disciplines outside of geography. A case study is described in which GST instruction is delivered through a study abroad program to the islands of Micronesia.

Keywords Higher education • Pedagogy • Trends • Micronesia

7.1 Introduction

Describing the role of Geospatial Technologies (GSTs) in higher education is not a simple task. The increased power, simplicity, and availability of GSTs have resulted in dramatic increases in applications, users, and instructors. In turn, these have sparked diverse pedagogical practices, objectives, and interdisciplinary approaches, involving both teaching *about* GSTs (i.e., emphasizing immediately applicable GSTs skills, knowledge, and experiences) and *with* GSTs (i.e., using GST concepts, products, or approaches to provide contexts or frameworks for non-GST content). This chapter, after considering the current major trends,

R. Perkins (✉)
Department of Environmental Science, Queens University of Charlotte, Charlotte, NC, USA
e-mail: perkinsr@queens.edu

O. Muñiz Solari et al. (eds.), *Geospatial Technologies and Geography Education in a Changing World*, Advances in Geographical and Environmental Sciences,
DOI 10.1007/978-4-431-55519-3_7

applications, and challenges of GSTs in higher education, presents a case study exemplifying one approach to teaching GSTs internationally.

7.2 Teaching *About* GSTs

The dominant paradigm in international higher education remains teaching *about* GSTs. This is largely due to the direct link between GSTs capacity and economic development, resource management, and infrastructure design, and the fact that developing a trained workforce (and employable graduates) is an explicit goal of higher education institutions. The U. S. Department of Labor (2005) identified GST as one of the top three economic growth sectors in the U. S.

Recognizing this importance, higher education institutions (often with governmental assistance) have developed content-driven GST curricula. In India, the pressing need for trained GST workers has led to increasing public-private educational partnerships to offer GST certificate, degree, and diploma programs in applied fields such as marine science, agriculture, and forestry (Tejpal et al. 2012). Similarly, Fuling and Shaohua (2007) argue that the first step to strengthening GIS instruction in China is to train students in the technical and operational details of spatial information science. In Brazil, Camboim and Brandalize (2013) describe the role of universities within the National Spatial Data Infrastructure (NSDI) project as preparing a qualified workforce, primarily by delivering a four-level capacity building curriculum involving increasingly complex GST capacities: information transmission, development of skills, development of attitudes, and development of concepts.

To facilitate GST instruction, many countries are developing national networks between higher education institutions. These not only help overcome local limitations of GST resources or expertise, they also standardize (and presumably strengthen) students' GST skills and knowledge. Importantly, they also extend the reach of GST instruction to a wider population. The 25 universities participating in Brazil's NSDI are creating a nationwide higher education network to develop GST curricula and distance learning techniques to reach more students (Camboim and Brandalize 2013). A network of U. S. community colleges, the GeoTech Center, exists explicitly to develop students' workplace GST skills, and to assist development of competency-based GST curricula in member institutions (Johnson 2012).

Many regional international higher education networks are also being developed for GST instruction. Examples include those in Africa, with the African Geospatial Sciences Institute (AGSI) (Oeldenberger and Khaled 2012), Europe, with the European Education in Geodetic Engineering, Cartography and Surveying project funded by the EU (Lisec and Fernandez 2008), and globally, with the GI-NET partnership developed by the International Institute for Geo-Information and Earth Observation (Geo-ITC) (Molenaar and Mannaerts 2008). These efforts are built on higher education's potential for trans-national education and research, the benefit to students of multiple sources of instruction and partnership, and web-based collaboration, including virtual campuses (e.g., Geo-ITC).

7.3 Teaching with GST

Increasingly, countries are complementing the approaches above by teaching *with* GST in higher education, thus including geographic and GST concepts, terminology, and techniques in courses and departments outside geography. The adaptability and utility of GSTs to so many applications, as Sinton (2011) notes, is one of its great strengths.

Two outcomes typify this trend. First, GST instruction and faculty are often housed in a variety of academic departments, from Geography to Environmental Science to Biology (Johnson and Sullivan 2010). This necessarily results in GSTs being used to advance the content of diverse curricula. Second, the use of GSTs in an ancillary or supportive role in non-GST courses is now established across the academic spectrum, from social sciences (Goodchil and Janelle 2010), to humanities (Bodenhamer et al. 2010), to business (Pick 2005), and even to art (Beech 2014). While the first example appears to be more common in the U. S., the second is decidedly global. Sanchez et al. (2010) have developed a GST-based role-playing game in which higher education students develop understanding of active citizenship in a hypothetical French town. In Japan, Yuda and Itoh (2006) argues that higher education faculty should strive to work collaboratively between disciplines and demonstrate the impact of GSTs in education, regardless of subject matter. This is particularly true as faculty teach pre-service K-12 teachers.

GST also has become ubiquitous in daily life. Higher education students potentially arrive in class having used GSTs to navigate their cars, geotag their latest selfie to Facebook, and notify fellow drivers of police car locations. This kind of "neogeography" (Turner 2006; Goodchild 2009) – the use of cartographic tools and techniques by non-expert users (both students and faculty) – shows that GST is no longer the purview of academia and the natural sciences.

Another factor contributing to GST's adoption across disciplines is its explicit development of (1) critical thinking and (2) spatial literacy, i.e., the ability to use the properties of space to communicate, reason, and solve problems (Hsin-Fu 2011). As the National Research Council (2006) states, "... without explicit attention to spatial literacy, we cannot meet our responsibility for equipping the next generation of students for the twenty-first century." GST also helps achieve the "essential learning outcomes" listed by AAC&U (2007) [quickly becoming the de facto standard in American higher education], including inquiry and analysis, critical and creative thinking, written and oral communication, quantitative and information literacy, and teamwork-based problem solving. However, GST is not always accepted as part of general education. Tsou and Yanow (2010) note that a lack of adequate facilities, skepticism amongst geography faculty, the high cost of data, and low public awareness of GSTs all deter use of GSTs in general education.

7.4 Best Pedagogical Practices

There is a widespread recognition amongst higher education instructors that teaching with and about GSTs is neither easy, nor intuitive. Numerous resource websites have been created (e.g., serc.carleton.edu, edcommunity.esri.com, www.teachgis.org) that provide not only lesson plans and pedagogical suggestions, but also de facto emotional support (e.g., the tagline of teachgis.org is "Because no one should have to face GIS alone."). The sites are clearly premised by the notion that teachers may feel overwhelmed by the potential of GSTs and a bit baffled by how to proceed. As Howarth and Sinton (2011) note, the two most influential statements on GIS curriculum development (i.e., the National Center for Geographic Information Analysis Core Curriculum in GIS (Goodchild and Kemp 1992) and the Body of Knowledge developed through the University Consortium of Geographic Information Science (DiBiase et al. 2006)) state *what* should be taught, but offer no guidance as to *how*.

Still, progress is being made in establishing effective pedagogical practice for higher education GST. Fuling and Shaohua (2007), in their argument for China's GIS education development, state that while emphasizing GST content is necessary, it is not sufficient to train innovative GIS talent. Faculty must complement students' technical GST knowledge with active participation in domestic and international exchanges, practical research projects, and collaborative partnerships with other students and with faculty. Aina (2012) found that his undergraduate students in Saudi Arabia learned better with approaches featuring active-learning with case studies, and "learning by teaching," in which students were responsible for presenting course content to each other. Schulz (2012) states that similar approaches work in the U.S. specifically as they affect student cognition and spatial reasoning skills.

These approaches are consistent with George Kuh's (2008) more general work identifying "high impact" pedagogical practices (i.e., those which enhance student engagement and increase student academic success) in universities. Specific examples applicable to higher education GST include (1) collaborative projects, in which students solve problems in the company of others and by listening to others' insights and experiences; (2) undergraduate research involving students with actively contested questions, empirical observation, cutting-edge technologies, and the sense of excitement that comes from working to answer important questions; (3) global learning, including domestic and study abroad programs helping students explore cultures, life experiences, and worldviews different from their own; and (4) community-based learning, requiring students to analyze and solve problems in the community. Regardless of the specific practice used, Favier and Van der Schee (2012) conclude that effective approaches in GST instruction require considerable investment by faculty to ensure students structure, correct, and expand their geographic thinking.

Finally, technological advances in recent years have been impressive and transformative for GST higher education. Cloud-based GSTs (e.g., ArcGIS Online,

Google Earth) and virtual globes (Songer 2011; Schulz et al. 2008), crowd-sourced data (or "volunteered geographic information") (Goodchild 2007), and the availability of GSTs on multiple platforms (e.g., desktops, tablets, and phones) continue to push the creativity of educators for new ways to teach GST.

7.5 Challenges and Prospects for the Future

Despite its great growth and successes, GST in global higher education still faces a number of challenges. First, while GST is becoming more commonplace in daily life, its availability in higher education is strikingly limited. Only 285 4-year institutions (of the more than 2800) in the U.S., offer BA or BS degrees in Geography (USDE National Center for Education Statistics 2014). For 2-year schools, only 451 of the 1184 total offer any type of geospatial education (courses, certificate, or degree programs) (Rudibaugh and Ferguson 2010). No data are available describing the number of U. S. schools that offer GST courses without a degree program, or teach with GST in non-geography courses. In India, only 111 higher education institutions (74 publics and 34 privates) offer GST courses of study (Tejpal et al. 2012). In the UK, the situation is proportionately better, with approximately 80 of the 160 degree-granting universities and colleges offering geography-related courses of study (Geography in the UK 2014). The limited availability of university-delivered GST programs in some countries is one reason for the increasing number of institutional networks and virtual campuses mentioned above.

Second, there is a deep concern that GST is neither attracting nor graduating a sufficient number of GST students. Mills et al. (2004) cite a worldwide decline in the early 2000s of the number of students beginning GST-related degree programs. Aina (2012), Tejpal et al. (2012), Camboim and Brandalize (2013), and McDougall et al. (2006) offer indication that this trend continues in Saudi Arabia, India, Brazil, Australia, and New Zealand, respectively. Partnerships between higher education institutions, industry, and government in each of these countries and the U. S. have been formed in part to increase student numbers. In at least one country, the UK, the total acceptances to undergraduate geography degrees have increased recently, doing so by 6.2 % between 2012 and 2013 (Geography in the UK 2014).

Perhaps the greatest challenge facing GST in higher education is the rate of change in GST itself (Unwin 2012). As instructors become familiar with existing GST products, software, and workflows, newer products and versions are released, thus making it difficult to remain current. Because of this, and the high number of GST instructors without formal GST training or home-institutional peer support, there will be a chronic need for formal and informal inter-institutional peer networks and collaborative learning opportunities for faculty. Programs in the U. S., such as Esri's Teachers Teaching Teachers GIS (i.e., T3G), and Integrated Geospatial Education Technology and Training (i.e., iGETT), are excellent examples of programs that have an explicit goal of fostering peer networks. This need for

GST instructors and institutions to develop formal and informal networks has been identified in Europe (Lisec and Fernandez 2008; Höhle 2006), Africa (Oeldenberger and Khaled 2012), China (Fuling and Shaohua 2007), and globally (Molenaar and Mannaerts 2008), and speaks to the role of higher education GST not only for student education, but also for faculty research and professional development.

Almost certainly, the future of GST in higher education will be marked by the continuation of the trends mentioned above: the integration of teaching *about* and *with* GSTs, the application of GSTs in an increasing range of academic disciplines, and in modes of delivery ranging from the casual to the highly quantitative and analytic with increasing use of technology to foster networks and collaborations, including those in a cloud-based learning environment. These shifts require dependable classroom and institution technological capacity (i.e., bandwidth, connectivity), but overcome limitations of hardware and software availability, capability, and currency. In addition, as many schools adopt wi-fi and tablet computers in the classroom, cloud-based applications make GSTs accessible without use of a specialized computer lab.

7.6 GST and GP in the Field: A Case Study in Micronesia

7.6.1 Introduction and Objectives

Queens University of Charlotte is a small (enrollment = 1300) private liberal arts university in Charlotte, NC, USA. One GST course (Introduction to GIS) is offered through the Department of Environmental Science, and there are no degree programs in Geography or GST. As a result, very few students are exposed to GST content or theory. However, Queens does have a robust international education program that ensures each student can study abroad. The following case describes an effort to integrate teaching about and with GST for Queens students on a study abroad trip to the islands of Yap (9.5 N, 138E), a member state of The Federated States of Micronesia (Fig. 7.1).

Like all small island developing states, Yap faces a long list of interconnected challenges, including climate change, invasive species, as well as a lack of accurate monitoring data, limited technological and educational resources. To help support resource management, planning, and external grant applications, there is a pronounced need for geospatial information (e.g., island physiography, infrastructure, demographics, resources, etc.) and analysis.

Unfortunately, despite the recognized need for local GST capacity, Yap has a limited ability to develop one (Perkins and Xiang 2006). Geography and GST are not included in the standard K-12 or local community college curricula. Teachers have no GST training, technology, or budget support. Beginning annually in 2001, Queens has taken students and faculty to Yap to conduct a range of GST-based

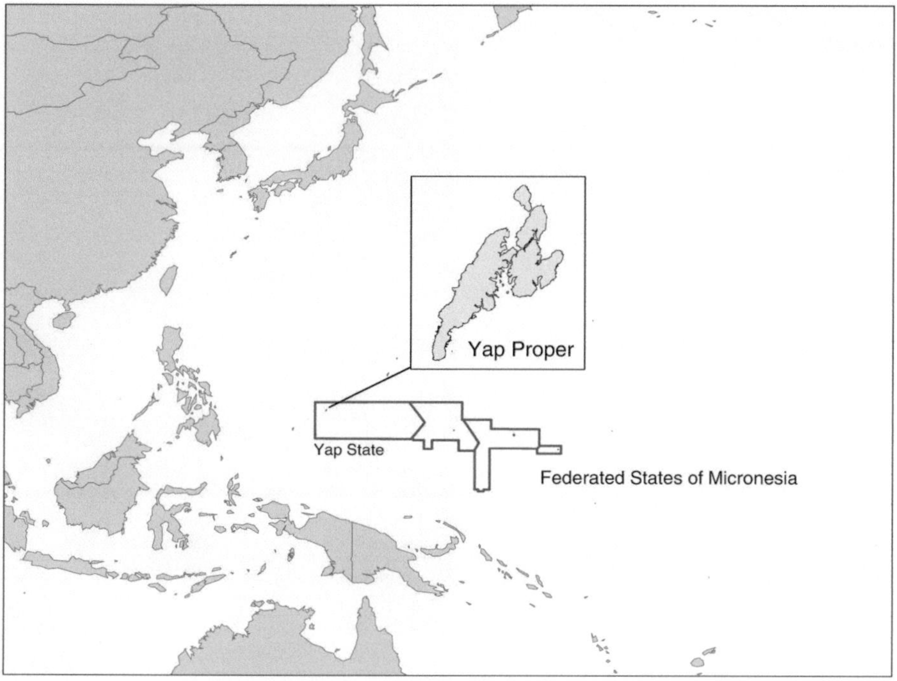

Fig. 7.1 Map of Micronesia with *inset* map of Yap Proper

resource management projects and simultaneously develop a local GST capacity. Through this approach, students learn through a problem-based approach and develop a keen sense of sound geospatial practice.

7.6.2 Preparation and Implementation

Prior to departure for Yap, students are required to complete a semester-long preparation course covering cultural, social, economic, historic aspects of Yap. ArcGIS Online is used both passively (as part of lectures to provide spatial context) and actively (as part of student exercises, e.g., to explore general and specific geographic questions, patterns, and trends related to Yap). In addition, the course covers the theory and practice of GSTs, in particular GPS and GIS. Typically, we GPS the major features of campus, then import the data into ArcGIS.

Once on Yap, the intended learning outcomes include a demonstration of progress in the application of GPS and GIS knowledge, critical thinking, oral communication, collaboration, and intercultural knowledge. As the specific work projects are identified, students are divided into small groups (typically 3–4 students) and matched with counterparts from the Yapese agency spearheading that project. Together, they use field GPS to map the desired features and make detailed

Fig. 7.2 Yapese counterpart and Queens student using field GPS

field sketches. The driving pedagogy relies on these high-relevance applications, the integrated delivery of theory and practice, and hands-on learning. Pedagogical practices of collaborative learning, undergraduate research, global learning, and community-based learning are delivered through this collaborative fieldwork.

Once field GPS data are collected, students differentially correct them using post-processing with data from a temporary GPS base station. The corrected data are examined and cleaned (i.e., corrected while comparing the digital version with hand-drawn field maps) before the final version is placed into the permanent geodatabase. Yapese counterparts keep a copy of all data and map products (Fig. 7.2).

7.6.3 Main Outcomes

7.6.3.1 Pedagogical

Students transition from a point of limited knowledge of GPS and GIS to a point of near-fluency by their field GPS work (i.e., operating in difficult environments, taking accurate field sketches) and lab GIS follow-up. Critical thinking is developed through assessing strategy for field data collection, troubleshooting the inevitable hiccups, and reconciling the actual data collected with their sketches. Oral communication skills are honed through the instructional process as students lead Yapese counterparts through fieldwork. This intercultural collaboration is critical to the educational success of the program. Importantly, teaching local counterparts about GST strengthens Queens students' skills and knowledge. This integration of problem-based learning and "learning by teaching" has proven to be incredibly effective.

7.6.3.2 Resource Management

While there have been over 20 projects conducted over the duration of the program, two examples demonstrate results from teaching *with* GST. The first project involves mapping the extent of a highly invasive (and highly resilient) grass, *Imperata cylindrica*, that had become established and was starting to spread via aerially dispersed seeds. The total affected area decreased from 14.5 ha in 2001 to less than 0.25 ha in 2007. Second, burned areas (i.e., both wild and anthropogenic grassland fires) have been mapped annually since 2001. The period-of-record and regular time-step of the data set are uncommon, if not unique, in Micronesia. As a result, the U. S. Forest Service used these data to build and validate a predictive fire model for Micronesia. In both examples, students learned the power of GST to further the fields of ecology, resource management, and computer modeling.

7.6.3.3 Capacity Building

The number of Yapese agencies actively using GST (i.e., data, software, and/or products) has increased from zero in 2001 to eight in 2014. An informal GST interest group has formed, and there has been a widespread acceptance of GST as a vital contribution to resource management and planning. Success has been due to two related factors: (1) the long-term nature of the Queens-Yap relationship, and (2) local control over the projects, the personnel, and the pace of progress. Building Yaps' GST capacity has occurred simultaneously with the increase in students' GST capacity. During their time on Yap, Queens students become the GST instructors: explaining the relevant GST theory and practice to their counterparts.

7.7 Conclusions

As GST evolves to become increasingly powerful, simple, and commonplace, it will continue to be relevant and vital within higher education. GST will contribute to not only geography and science-related disciplines, but also general education and workplace skill development. Many countries' initial strategy of teaching *about* GST in higher education will continue to be complemented by efforts to teach *with* GST in non-geography departments and courses. Technological developments in web or cloud-based delivery of both GST content and analytical capacity are increasing the potential for networked relationships both at the institutional and peer levels. In turn, these new opportunities are spurring new efforts to expand the impact of GST instruction to new areas and students.

Nevertheless, there are challenges facing higher education GST instruction. It is likely that the vast majority of higher education students will not encounter GST through a formal GST course, certificate, or degree program. The majority of 2- and 4-year colleges and universities do not offer courses, certificates, or degrees in either geography or GST-based areas of study. Instead, GST will be more likely presented as a supporting or ancillary device in non-science courses, or through courses in spatial thinking or reasoning. GST, originally created as an analytical tool for geoscientific research, increasingly will be presented to students by instructors who themselves likely have not had any formal training in GST outside a workshop or short course. In addition, the pace of change of software, language, as well as GST's curricular role, makes it difficult for instructors (regardless of their original training) to maintain currency in GST knowledge and skill. Finally, the varied curricular role of GST (i.e., for cognitive, workplace, and disciplinary skills), will almost certainly require different pedagogical approaches. It seems reasonable to borrow from more general literature and use pedagogies (e.g., Kuh 2008) to make GST relevant to the student and engage students in activity-based and collaborative projects.

One approach to higher education GST instruction is the Micronesia Program at Queens University of Charlotte. This program employs a range of pedagogical techniques, including problem-based learning, collaborative learning, learning by teaching, and community-based learning within a study abroad experience. Because GST is presented to students as seamlessly integrated with environmental policy, monitoring, and management, students improve their knowledge of GST and why it is critical to environmental conservation. This pedagogical approach of integrating GST instruction with long-term place-based monitoring can be applied in any location.

References

AAC&U. (2007). *College learning for the new global century*. Washington, DC: Association of American Colleges and Universities.

Aina, Y. (2012). Applications of geospatial technologies for practitioners: An emerging perspective of geospatial education. In *Emerging informatics – Innovative concepts and applications* (pp. 1–20). Rijeka: InTech.

Beech, D. (2014). *GIS as an artistic medium*. http://www.academia.edu/2574558/GIS_as_an_Artistic_Medium. Accessed 18 July 2014.

Bodenhamer, D., Corrigan, J., & Harris, T. (Eds.). (2010). *The spatial humanities: GIS and the future of humanities scholarship*. Bloomington: Indiana University Press.

Camboim, S. P., & Brandalize, M. C. B. (2013). *The role of universities in the Brazilian NSDI capacity building plan – The case of the open geospatial laboratory at UFPR*. International

Cartographic Association. http://icaci.org/files/ documents/ICC_proceedings/ICC2013/_extendedAbstract/430_proceeding.pdf. Accessed 25 Aug 2014.

DiBiase, D., DeMers, M., Johnson, A., Kemp, K., Luck, A. T., Plewe, B., & Wentz, E. (Eds.). (2006). *Geographic information science and technology body of knowledge*. Washington, DC: Association of American Geographers.

Favier, T. T., & Van der Schee, J. A. (2012). Exploring the characteristics of an optimal design for inquiry-based geography education with geographic information systems. *Computers & Education, 58*(1), 666–677.

Fuling, B. I. A. N., & Shaohua, W. (2007). Problem, innovation and development of GIS higher education in our country [J]. *Geomatics World, 2*, 003.

Goodchild, M. (2007). Citizens as sensors: The world of volunteered geography. *GeoJournal, 69*, 211–221.

Goodchild, M. (2009). NeoGeography and the nature of geographic expertise. *Journal of Location Based Services, 3*(2), 82–96.

Goodchild, M., & Janelle, D. (2010). Toward critical spatial thinking in the social sciences and humanities. *GeoJournal, 75*(1), 3–13.

Goodchild, M., & Kemp, K. (Eds.). (1992). *NCGIA core curriculum: Application issues in GIS* (Vol. 3). Santa Barbara: National Center for Geographic Information and Analysis, University of California at Santa Barbara.

Höhle, J. (2006). Education and training in photogrammetry and related fields–remarks on the presence and the future. *International Archives of Photogrammetry, Remote Sensing and Spatial Information Sciences, Commission VI, 9*.

Howarth, J., & Sinton, D. (2011). Sequencing spatial concepts in problem-based GIS instruction. *Procedia-Social and Behavioral Sciences, 21*, 253–259.

Hsin-Fu, W. (2011). *Why spatial literacy?* Carleton College. https://apps.carleton.edu/collab/spatial_analysis/SpatialLiteracy/. Accessed 18 July 2014.

Johnson, A. (2012). Geospatial education at U.S. community colleges. In *Teaching geographic information science and technology in higher education* (pp. 185–198). West Sussex: Wiley-Blackwell.

Johnson, A., & Sullivan, D. (2010). Geospatial education at U. S. community colleges: Background, challenges, and opportunities. *URISA Journal, 22*(2), 5–11.

Kuh, G. (2008). High-impact educational practices: What they are. *Who has access to them, and why they matter, Association of American Colleges and Universities*. http://accreditation.ncsu.edu/sites/accreditation.ncsu.edu/files/Kuh_HighImpactActivities.pdf. Accessed 25 Aug 2014.

Lisec, A., & Ruiz Fernández, L. Á. (2008). *European project on higher education in the fields related to geomatics as support for mobility of students and teachers*. International Society for Photogrammetry and Remote Sensing. http://www.isprs.org/proceedings/XXXVII/congress/6a_pdf/5_WG-VI-5/05.pdf. Accessed 25 Aug 2014.

McDougall, K., Williamson, I., Bellman, C., & Rizos, C. (2006). Challenges facing spatial information and geomatics education in the higher education sector. In *Proceedings of the combined 5th trans Tasman survey conference and 2nd Queensland spatial industry conference 2006: Land and sea spatially connected-in a tropical hub*. Spatial Sciences Institute, Canberra, Australia.

Mills, J. P., Parker, D., & Edwards, S. J. (2004). Geomatics. org. uk: A UK response to a global awareness problem. *International Archives of the Photogrammetry, Remote Sensing and Spatial Information Sciences, 35*(B6), 1–6.

Molenaar, M., & Mannaerts, C. (2008). Capacity building networks for earth observation and GEO-ICT a first step towards a Virtual University. In *ISPRS 2008: Proceedings of the XXI congress: Silk road for information from imagery: The international society for photogrammetry and remote sensing, 3–11 July, Beijing, China. Comm. VI, WG VI/3* (pp. 93–96). Beijing: ISPRS.

National Center for Education Statistics. (2014). *College navigator*. National Center for Education Statistics. http://nces.ed.gov/. Accessed 18 July 2014.

National Research Council. (2006). *Learning to think spatially*. Washington, DC: National Academy Press.

Oeldenberger, S., & Khaled, K. B. (2012). The African Geospatial Sciences Institute (agsi): A new approach to geospatial training in North Africa. *ISPRS-International Archives of the Photogrammetry, Remote Sensing and Spatial Information Sciences, 1*, 45–51.

Perkins, R. M., & Xiang, W. N. (2006). Building a geographic info-structure for sustainable development planning on a small island developing state. *Landscape and Urban Planning, 78* (4), 353–361.

Pick, J. B. (Ed.). (2005). *Geographic information systems in business*. Igi Global.

Royal Geographical Society. (2014). *Geography in the UK*. Royal Geographical Society. http://www.rgs.org/GeographyToday/GeographyintheUK.htm. Accessed 16 Sept 2014.

Rudibaugh, M., & Ferguson, B. (2010). Mapping geospatial education at U. S. community and technical colleges. *URISA Journal, 22*(2), 41–44.

Sanchez, E., Delorme, L., Jouneau-Sion, C., & Prat, A. (2010). Designing a pretend game with geotechnologies: Toward active citizenship. *Learning with Geoinformation, V*, 31–40.

Schulz, R. (2012). Active pedagogy leading to deeper learning: Fostering metacognition and infusing active learning into the GIS&T classroom. In *Teaching geographic information science and technology in higher education* (pp. 133–144). West Sussex: Wiley-Blackwell.

Schulz, R., Kerski, J., & Patterson, T. (2008). The use of virtual globes as a spatial teaching tool with suggestions for metadata standards. *Journal of Geography, 107*(1), 27–34.

Sinton, D. (2011). Making the case for GIS in higher education. In *Teaching geographic information science and technology in higher education* (pp. 17–36). West Sussex: Wiley-Blackwell.

Songer, L. (2011). Using web-based GIS and virtual globes in undergraduate education. In *Teaching geographic information science and technology in higher education* (pp. 289–300). West Sussex: Wiley-Blackwell.

Tejpal, M., Jaglan, M. S., & Sharma, K. (2012). Significance of geospatial technology and status of its education and training in India. *International Journal of Geomatics and Geosciences, 2*(3), 731–737.

Tsou, M. H., & Yanow, K. (2010). Enhancing general education with geographic information science and spatial literacy. *URISA Journal, 22*(2), 45.

Turner, A. (2006). *Introduction to neogeography*. Sebastopol: O'Reilly Media.

Unwin, D. (2012). Discussion and prospect. In *Teaching geographic information science and technology in higher education* (p. 247). West Sussex: Wiley-Blackwell.

U.S. Department of Labor. (2005). *High growth job training initiative*. Geospatial Technology Apprenticeship Program. http://www.doleta.gov/BRG/JobTrainInitiative/. Accessed 18 July 2014.

Yuda, M., & Itoh, S. (2006). Utilization of geographic information systems in education reform in Japan. In *9th AGILE international conference on geographic information science. "Shaping the future of geographic information science in Europe", Thermal Hotel, Visegrád, Hungary* (pp. 20–22).

Chapter 8
Practice of Geospatial Technologies in Informal Learning

Osvaldo Muñiz Solari and Melody Crenshaw

Abstract This chapter focuses on the most recurrent methods used by practitioners to practice geospatial technologies (GSTs) in informal learning.

GSTs are used by a wide range of practitioners. Being either new young generations acting as digital natives or adults with professional experience, all of them are engaged with new technologies in one way or another. If geospatial practice (GP) is defined as organized activities to analyze and interpret geospatial phenomena by these practitioners, a great deal of these exercises can be attributed to non-formal practices engaged through informal education.

The fast and transformative expression of crowdsourcing as well as volunteered GST has given new positions to informal education. Traditional roads to acquire knowledge are challenged by new ways to practice geospatial visualization to resolve Earth's problems in terms of physical and human transformation.

Voluntary participation, intense engagement with global tasks, individual and group improvisation, and flexible innovation are among the methods that are analyzed and discussed in the construction of GP. Ultimately, the free-choice learning is put in perspective to verify how effective informal education is to tackle complex geospatial problems that are global in nature.

Keywords Informal learning • Geospatial practice • Crowdsourcing

8.1 Introduction

The continuous improvement of information communication technologies (ICTs) through Internet has created a new global perspective to deal with complex geographic phenomena in our planet. It goes without saying that language barriers,

O. Muñiz Solari (✉)
Department of Geography, Texas State University, San Marcos, TX 78666, USA
e-mail: o.muniz@txstate.edu

M. Crenshaw
Northwest Vista College, San Antonio, TX 78251, USA
e-mail: melody.crenshaw@gmail.com

O. Muñiz Solari et al. (eds.), *Geospatial Technologies and Geography Education in a Changing World*, Advances in Geographical and Environmental Sciences,
DOI 10.1007/978-4-431-55519-3_8

cultural differences and religious divergences create clear obstacles for interaction and even collaboration, yet millions of people from different countries and continents are willing to act very dynamically by using ICTs to resolve problems. It is the presence of a new and wide-reaching form of learning that crosses all levels of the global society. Not being new, informal learning (IL) has been transformed into a powerful mechanism for self-independence. It has allowed people to reach new forms of power practiced through self-learning activities. Self-learning education has been one of the great benefits that people around the world have experienced with the development of ICTs. Among the ICTs, geospatial technologies (GSTs) epitomize the essence of tools and mechanisms to learn about our planet and how to improve the spatial conditions and information that result from it.

As authors of this chapter, we first clarify the difference between IL and non-formal learning, then discuss the importance of informal learning about geospatial technology (GST) and with GST. After that, the geo-enabling world and its effects on IL are also uncovered for examination. Finally, the introduction of a global example represented by volunteered geographic information (VGI) pretends to show the power of IL. The practice with GST to resolve geographic problems of global dimensions is here represented in its more contemporary expression.

8.2 Informal and Non-formal Learning: The Need for Clarification

We all have a clear understanding about what formal education is. When education is extended outside of the systematic procedures of teaching and learning, which also include distance and online education, we need to understand other forms of acquiring knowledge. Then it is very important to make a clear distinction between non-formal education and IL; two concepts that create confusion and different positions. However, the Organization for Economic Co-operation and Development (OECD) has recognized the importance of IL and non-formal learning represents a rich source of capital in any society that value the importance of all form of knowledge acquisition (Werquin 2010).

Let us explain our position about these two concepts. For that purpose, we accept Rogers' paradigm about non-formal education, and IL. He points out:

> Informal learning here, being all that incidental learning, unstructured, unpurposeful but the most extensive and most important part of all the learning that all of us do every day of our lives, as I have shown elsewhere. (Rogers 2003)

Later on, he also clarifies by saying:

> There are not of course categories. The boundaries between each of these 'sectors' are very fuzzy indeed. But the distinction is very real. Learning is the keystone; it is the original matter out of which all education is created. Somewhere along the learning continuum, we

> come to purposeful and assisted learning (education in its widest sense). When we control this and individualise it, learn what we want for as long as we want and stop when we want, we are engaging in informal education. When we step into a pre-existing learning programme but mould it to our circumstances, we are engaged in non-formal education. When we surrender our autonomy and join a programme and accept its externally imposed discipline, we are immersed in formal education. (Rogers 2004)

That said, we consider any education through special activities out of the school such as visiting a museum or doing field work as non-formal education. Both are, in some way guided by special conditions. When children visit a museum and especially when they are guided or accompanied by teachers they are directed through some rooms, develop some activities, and read special materials on display. Usually, children are confronted with problems and questions that are displayed on purpose to guide the visitors. By the same token, when children go to the field with teachers they are also guided and normally following some steps or guidance to comply with some rules and reach some objectives. Both examples represent non-formal education. They are not IL because the latter is incidental learning and unstructured. The one we develop every single day. We learn what we want and we can stop when we want. Those conditions are not present when children visit a museum or develop field work.

Our chapter is not only about young people (school and non-school students) but also adults who are professionals or non-professionals. However, given the fact that we are dealing with GST it normally requires some expertise. Consequently, we expect to find adults practicing these technologies more often, yet we have to recognize that young people; even at the level of middle and high school, are increasingly using some of these technologies. Therefore, we are also interested in the process of practicing GST by youngsters and the way they learn how to improve skills outside of their school environment.

8.3 Informal Learning: About GST and Resolving with GST

Learning about GST and resolving geospatial problems with GST are initially accomplished via formal education. Nevertheless, the increasing influence of available resources in the Web 2.0 has put a variety of software, data, and apps in public hands to acquire knowledge about GST and develop further practice with GST through IL. The primary effect of this trend is characterized by a global dimension of practitioners who, as users and producers, are generating a growing network of volunteers that Budhathoki et al. (2008) have conceptualized as "produsers". Hidden, as it is the complex network of helpers, is the important process of IL that we want to examine.

8.3.1 *Away from Pedagogy and Near Andragogy. IL and GST*

While recognizing the strength of most of the arguments that pedagogy is derived from the Greek meaning literally, 'child-leading', and by the same token, andragogy is recognized as 'man-leading', there is the assumption that, in the process of learning, pedagogy is focused on children and andragogy is associated with adults. However, the intention here is not the initiation of a semantic conversation but the interest for understanding the continuous learning process. It is a progression initiated by formal guidance that regenerates in any person's life time through particular learning with minimum or rather absent supervision. This latter condition is an important characteristic of a self-education process that individuals are exposed to as they are increasingly involved with ICT.

The interpretation of andragogy by Knowles (1980) as a learning process that finally focuses on the internal motivation by a person to learn as he or she matures, brings about the importance of independence and self-direction in the progression of discovering. Instead of a closed system that portrays a pedagogical approach in formal education, Muñiz-Solari (2014) finds a rising consensus of a wide open system available to people of all ages willing to acquire knowledge through informal learning (Fig. 8.1).

The evidence that a global trend is taking place which shows a growing number of people engaged in a wide range of technology-based IL (Cranmer 2006; Jones and Conceição 2008; Jamieson 2009; Goodwin et al. 2010) confirms the importance of social networks. Farnham et al. (2012), citing Ito (2010) point out the

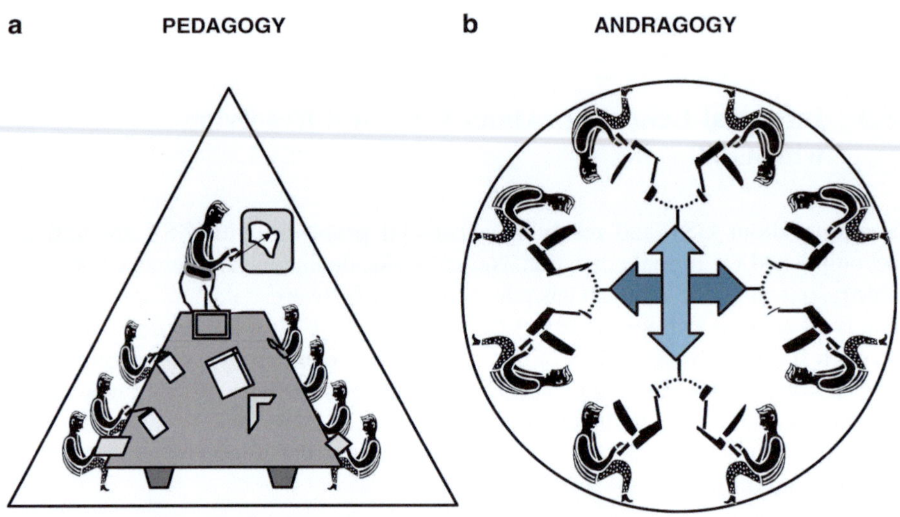

Fig. 8.1 The closed system of formal education (**a**) normally conducted in a room and the open system of informal learning (**b**) developed in the geo-enabling world (Muñiz-Solari 2014)

significance of interest-based networks with peer-based sharing which are conducive to IL. Through interaction with their peers, youngsters get away from the closed system of a pedagogical environment and reach closer to the real life of adulthood. Nevertheless, interest-based networks do not accomplish a higher level of collaboration and complexity. Among these networks is the one that involves a multitude of amateurs, professionals, and experts using GSTs. Consequently, the andragogical environment is reaching its stage of real configuration. Global teams, defined as global citizens engaged in IL take mutual responsibility to obtain common objectives.

The geo-enabling world has accelerated the process of networking where nodes, represented by organizations as well as individuals, cooperate and collaborate to gather data, build information, and resolve geographical problems. Open sources, open infrastructure, and open decisions are vital to operate with GSTs via IL.

8.3.2 Open GST Infrastructure and the Power of Open Data

With the development of Web 2.0 there has been a worldwide expansion of tools, and mechanisms to share and use GSTs on the Internet. Consequently, the process of IL about GST and with GST has been boosted among millions of practitioners; youth and adult global citizens who are eager for interaction and willing to capture geospatial information. The emergence of Web GIS is a classic example of the IL process that took place on the Web. First, the public learned from a Web-based map viewer and interacted inside a Web browser. After the first encounter with online exercises the IL practiced by amateurs and professionals experienced a tremendous change with Web 2.0.

The spatial abilities and the geospatial integrative skills developed by practitioners around the world have been increasingly enhanced by Web 2.0. This initial read-write Web has now progressed from Web-as information-source architecture and Web-as customer-self-service to Web-as participation-on multiple platforms (O'Reilly 2005; Prandini and Ramilli 2012). The Web as a platform allows the use of Software-as-a-Service (SaaS). GIS applications and the integration of GSTs (i.e., GIS, GPS, RM) and access to more devices create an enhanced process of usability, reliability and scalability, all delivered "in the cloud".

Geodetic control, orthoimagery, elevation, transportation, hydrography, governmental units, and the cadaster are the typical themes of the geographic data framework. The IL practiced by global citizens normally covers themes that do not demand high cost or high level of expertise. Elwood et al. (2012) point out that volunteers' collaboration to collect data and use them as main source of information are more related to transportation and hydrography. Through formal or informal gatherings global citizens compile geographic information to produce useful maps in a matter of days. Certainly, the IL process is part of this endeavor that has local or regional objectives, yet it is organized by large group of people of global dimensions. The power of open data is firmly rooted in the power of global citizens thanks to the creation of a geo-enabling world.

8.4 The Effect of the Geo-enabling World on IL

Users around the globe participate in generating value as they collaborate and resolve problems. We are in front of a Collective Intelligence, mentioned by Lévy (1999) or "crowd wisdom" (Surowiecki 2004), as a form of universally distributed knowledge, constantly enriched, coordinated in real time, and resulting in the effective mobilization of skills through the cyberspace. Ultimately, the whole process of collaboration is based on an IL initiative made by individuals and groups of people.

Without formal condition of learning, given by curricula and teachers, informal learners take their own choice both about what they like to learn and which GSTs they will select to make their contributions. The existence of a geo-enabling world might transform the way informal learners determine their decisions to acquire knowledge and what GSTs are more useful to initiate the IL process.

8.4.1 From Public Participation GIS Community to Community Integration GIS

One of the GSTs introduced to the world community very early during the expansion of the Internet was GIS. Web GIS, later on, created an increasing and ongoing participation. Public Participation GIS (PPGIS) encourages communities of digital amateurs and professionals to integrate local knowledge and resolve regional problems (Craig et al. 2002). PPGIS develops an IL process among members of local communities that empower them to enforce development planning processes.

Examples of PPGIS that take advantage of GST are numerous; all of which strive to integrate populations through geographic technology. Collaborative GIS joint venture in rural Australia (Walker et al. 2003), forest management through PPGIS application in Southern Ghana (Kwaku Kyem 2003), GIS for community forestry in Nepal (Jordan 2003), Web-based PPGIS in the United Kingdom (Kingston 2003), putting marginalized communities on the map with GPS and GIS technologies in Kibera slum, Nairobi, Kenya (Pétursdóttir 2011), and mapping for conservation and development in the Amazon borderland (Salisbury 2011) are, in fact, GST IL projects organized to accomplish the solution of local problems through collaboration. The latter example is a good representation not only of amateurs, professionals, and experts working together in non-formal and IL with GST, but also allowing the public (indigenous people in the Amazonia of Peru and Brazil) to practice IL with GIS maps.

PPGIS still has some pitfalls in the form of obstruction of the IL process because of external control on the final decision making process. Consequently, IL does not have the connotation of free-choice learning and flexible knowledge. Community-integration GIS has been the contemporary response to the aforementioned drawback. Community-Integrated GIS for Land Reform in South Africa (Weiner and Harris 2003), is an excellent example of what means to have free-choice learning and flexible knowledge. The authors suggest, however, that their experience in South Africa is not unique and that participatory GIS projects will be grounded in place-based fieldwork, keeping in mind local politics and institutional capacity. The redefinition of local institutions should achieve greater public participation and more efficient informal learning. Democratizing spatial decision-making is part of a critical unavoidable process to promote and encourage IL within the community.

8.4.2 Voluntary Participation: Citizens of World Through IL

We need to be prepared for a new world that is digital in its ability to interpret any earth phenomena and global in its challenges and treatments of worldwide issues. It is about the solution and resolution of local and regional geospatial problems that ignite connections with the rest of the world through a collective intelligence. Thus, we agree that the largest boost to GST in education is based on citizen science initiatives. Volunteered Geographic Information (VGI), as proposed by Goodchild (2007), is a growing trend of user-generated geographic content by thousands of individuals working in the Web 2.0 and the NeoGeographers who, for a lack of a better concept, Goodchild (2009) identifies as non-geographers producing geographic information because geography is experienced by everyone.

Even when VGI has proved to be efficient during emergencies, especially when natural disaster occur, Elwood et al. (2012) have questioned the quality of information produced by VGI which have to be taken with some care. Regarding the ability of the non-expert to integrate information as VGI and the quality of the knowledge being obtained, there is an unquestionable process of IL being built that is a result of social practice. If we also agree with the researchers that the VGI phenomenon has taken the production of geographic information away from the exclusive hands of geographers, the process of IL about GST and with GST has to be recognized as part of the world society. Examples of this trend is VGI based on crowdsourcing platforms in archeology that Sylaiou et al. (2013) introduces us with crowd collaborative works developed to combat elicit trafficking of antiquities (https://heritage.crowdmap.com) or crowdsourcing a project to identify archeological treasures in Mongolia (http://exploration.nationalgeographic.com/mongolia). On the other hand, business professionals have also proposed the Geographic Knowledge Discovery (GDK) as a process to uncover information in the large amount of digital geo-referenced data being collected, archived and shared by people using crowdsourcing platforms (Lee and Torpelund-Bruin 2011).

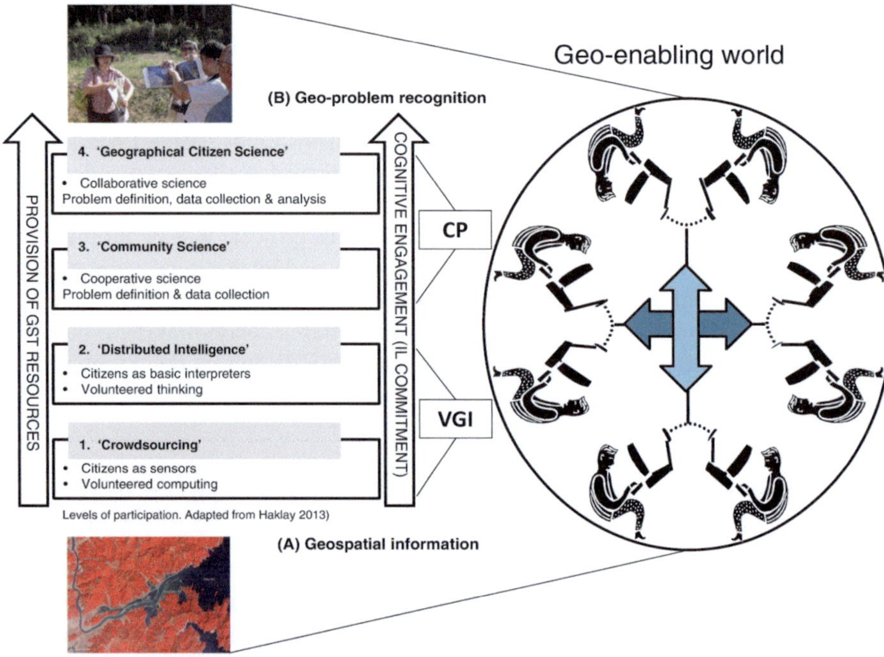

Fig. 8.2 The open system of informal learning developed in the geo-enabling world: Levels of participation from VGI to CP (Muñiz-Solari 2014)

Crowdsourcing platforms have allowed amateurs, professionals, and scientists to participate in local and worldwide projects. Some more than others, but eventually all participants go through a process of IL about GST and with GST. However, there is need for guiding the process of data quality and information reliability. Geography educators and researchers in Geography Education should respond quickly to this challenge by creating specialized communities of practice (CPs) to guide the processes of geo-problem recognition (Muñiz-Solari 2014). Consequently, IL in geography education could be monitored; giving geographers the opportunity to lead the process of local awareness on multiple geographic problems that volunteers are seeking to report. Muñiz-Solari (2014) shows the sequential process of participation that goes from gathering geospatial information to engaging in geo-problem recognition and ultimate solutions. Volunteers working at the different levels get not only an increasing provision of GST resources but also an increasing cognitive involvement in IL (Fig. 8.2). Therefore, unlike Haklay's hierarchy (2013), there should be a final and superior level of participation with GST known as 'Geographical Citizen Science'. This level of participation in which IL is at the highest expression is more than just the collection of location information as an integral part of the activity, as Haklay points out. It is the articulation of geo-problem recognition that encompasses multiple geographic components generating geographic patterns.

8.5 A Global Example of IL. Toward Citizen Science

As geospatial information is more available so is the number of amateurs looking for the more precise geographic information or producing it, as in the cases of crowd collaborative communities. Nevertheless, the technological setting in which especially amateurs are positioned (i.e., learning support system, platform, technical conditions) may create better or worse conditions to IL and; therefore, motivation to build knowledge in the Web. Within volunteer functions inventory, Budhathoki (2010) places understanding as an important one that represents the desire to learn.

IL processes within large groups of global citizens who deal with VGI falls under the concept of citizen science initiative. Initially defined, citizen science are scientific activities in which non-professional scientists voluntarily participate to collect data that follow some protocol. New conceptions of citizen science have generated an increasing development of IL. A specific type of activities termed ‘Geographical Citizen Science’ (Haklay 2013), brings IL to a new level of progress.

8.5.1 The Open Street Map and Some Derived Applications

OpenStreetMap (OSM) was developed in the UK as an effort to create a mapping software that was free to use and free from restrictive and proprietary copyrights, and therefore available to more users around the world. The data is VGI, and can be edited and analyzed by registered users. Although there are some gaps in data from lower socioeconomic areas where people are less likely to have GPS devices or, perhaps interest, the data that has been imputed by users is quite accurate (Haklay 2010). The contributors to OSM are relatively high-level, *Expert Authority, Expert Professional*, and even *Expert Amateur* (Coleman et al. 2009). *Interested Amateurs* might upload GPS recorded tracks, but *Neophytes* are unlikely to follow the steps necessary to volunteer information. Thus, OSM is a GST that is contributing to the IL of very interested community members. This community includes a Humanitarian OpenStreetMap Team (HOT) that is working to link responders in disaster areas with the OSM community, both adding local information from volunteers, but also providing necessary geographic information to the responders (http://hot.openstreetmap.org/about).

Looking for the main causes that move people to contribute and work with GST in informal settings took us to some of the most recent research findings. Budhathoki (2010) explains that learning new technologies, self-expression, and recreation are among the most important causes for IL. Furthermore, Neis et al.’s research (2013) on 12 selected urban areas; one for each continent, found that about 7 % of the data contributors are very active “Senior Mappers” while 28 % fall into the “Junior Mapper” category, and the largest group was classified as “Nonrecurring Mappers” with 66 %. Questions remain about possible causes to engage in

IL. Among those are: Internet access, culture, mentality, personal interests, and language barriers.

The IL that millions of people experience when they volunteer as communities of neogeographers around the world seems to emphasize Jonassen et al.'s five attributes of meaningful IL with technology (Jonassen et al. 2003; Howland et al. 2012). Active, constructive, cooperative, intentional, and authentic learning opportunities might be present in IL about GST and with GST. Clough (2010), when referring to Jonassen et al. in her paper on Geocaching, points out "...that although learning opportunities were integral to community membership, learning was not the main reason people joined the community or started Geocaching" (p. 33).

But the majority realized that IL was an outcome of membership. Geocaching members evolve through having engaging location-based experiences and creating new experiences for others. The community is growing in number and in participation as members contribute more hidden caches.

8.5.2 Citizen Science and 'Citizen as Sensors'. Rules and Responsibilities in IL

Much of the VGI contributed by non-professionals using GST is termed citizen science as that information is used to improve the knowledge and/or understanding about locations or phenomena associated with locations and places (Haklay 2013). Citizen science can provide vital details that researchers alone would take much longer to gather. In some cases, without the help of VGI, there would be more gaps in the research areas and overall understanding of a phenomenon. But not all volunteers contribute good information and certainly some of the weakness is unintended. There are the potential for dangerous intentions and false information too. So should there be rules and responsibilities about gathering data and information via IL and how and by whom should these be set?

Contributions to volunteer sites can be categorized as constructive or damaging (Coleman 2010, Coleman et al. 2009). Some of the reasons for constructive contributions he includes, among others, are: *Altruism, Professional or Personal Interest, Intellectual Stimulation, Social Reward, Pride of Place, Constructive Amendments or Clarifications, and Minor Edits and Format Changes.* Damaging reasons include: *Nonsense (text that is meaningless to the reader or irrelevant to the context of the article), Mischief, Agendas (social, economic, or political), Malice, Spam, and Partial or Mass Deletes.* Companies and organizations may establish levels of rules for contributors that may, in part, be based on the level of oversight they can provide, and the security they need, and the audience they serve (Coleman et al. 2009).

Various sources have implemented different levels of security for VGI. Some sites run by organizations may have a team that oversees the VGI process and will require registration with a legitimate email address before allowing uploaded data, but that data may need to fit certain parameters (a carefully scripted data entry screen for bird counts, for example) for acceptance. So, while data numbers have the potential to be wrong, the entry may be restricted to numbers, not text. Other sites such as OpenStreetMap may have less strict parameters and rely on community members to verify and/or edit new content. In many cases, edits cannot be made by other members, but the protocol requires a dialogue between the original data contributor and the individual making the claim of error.

The process of informal learning is also extended to rules and responsibilities. An increasing effect of monitoring is taking place among volunteers of environmental geographic information. The role of volunteers in support of crisis management is mentioned by Schade et al. (2010) referring to events analyzed by various researchers: De Rubeis et al. (2009) in earthquakes, De Longueville et al. (2009) in forest fires, Hughes and Palen (2009) in hurricanes, and De Longueville et al. (2010) in floods. Schade et al. contribute to the emerging field of VGI Sensing introduced by the latter researchers. Utilizing the concept of 'citizens as sensors' for sensitive data as the one produced from crisis events, Schade et al. propose not the citizen as a sensor making measurements, but as an element of a (virtual) VGI sensor. This includes statistically processing big amount of VGI to derive knowledge the same way as a remote sensing image is processed to translate spectral signature into geospatial knowledge. Consequently, the process of checking quality from geographic information produced by volunteers in their IL activities is similar to assessment procedures applied to youngsters in schools as part of their formal learning.

8.6 Conclusions

The increasing availability of GST infrastructure, data, and platforms in the Web 2.0 has allowed the citizens of the world to acquire knowledge about GST and develop further practice with GSTs through IL. This worldwide transformation of available knowledge in Internet has also reduced the difference between pedagogy and andragogy. Young people prepared through formal education are much closer to adult population through ICT than in the past, thus creating a bridge that connect population of all ages in their interest to better understand how to use local space as well as foreign environment.

The geo-enabling world has accelerated the process of networking between individuals and public and private organizations. A new citizen is formed who as either digital natives or digital migrants participate in generating value as they collaborate and resolve problems. The existence of this geo-enabling world might transform the way informal learners determine their decisions to acquire knowledge and what GSTs are more useful to use and integrate to develop the IL process.

As citizen of the world become more involved in the process of looking for efficient ways to occupy and use the space they inhabit, PPGIS encourages communities of digital amateurs and professionals to integrate local knowledge and resolve regional problems. Public participation paves the way to a growing trend characterized by global volunteers. VGI is formed as a result of digital collaboration to capture data and create massive information.

Despite the fact that crowdsourcing has moved millions of non-experts to integrate information as VGI, creating some question about the quality of information digitally gathered, there is an unquestionable process of IL being built that is a result of social practice. If we also agree with the researchers that the VGI phenomenon has taken the production of geographic information away from the exclusive hands of geographers, the process of IL about GST and with GST has to be recognized as part of the world society.

References

Budhathoki, N. R. (2010). *Participants' motivations to contribute geographic information in an online community*. Unpublished doctoral dissertation, University of Illinois at Urbana-Champaign, Champaign.

Budhathoki, N. R., Bruce, B., & Nedovic-Budic, Z. (2008). Reconceptualizing the role of the users of spatial data infrastructure. *GeoJournal, 72*(3–4), 149–160.

Clough, G. (2010). Geolearners: Location-based informal learning with mobile and social technologies. *IEEE Transactions on Learning Technologies, 3*(1), 33–44.

Coleman, D. (2010). Voluntereeded geographic informatioon in spatial data infrastructure: An early look at opportunities and constraints. In A. Rajabifard, J. Crompvoets, M. kanantari & B. Kok (Eds.), *Spatially enabling society: Research, emerging trends and critical assessment*. Leuven: Leuven University Press.

Coleman, D. J., Georgiadou, Y., & Labonte, J. (2009). Volunteered geographic information: The nature and motivation of produsers. *International Journal of Spatial Data Infrastructures Research, 4*, 332–358.

Craig, W. H., Harris, T. M., & Weiner, D. (Eds.). (2002). *Community participation and geographic information systems*. London: Taylor and Francis.

Cranmer, S. (2006). Children and young people's uses of the Internet for homework. *Learning Media and Technology, 31*(3), 301–315.

De Longueville, B., Smith, R. S., & Luraschi, G. (2009). "OMG, from here, I can see the flames!": A use case of mining location based social networks to acquire spatio-temporal data on forest fires. In *Proceedings of the 2009 international workshop on location based social networks* (pp. 72–80). Seattle: ACM Press.

De Longueville, B., Luraschi, G., Smits, P., Peedell, S., & De Groeve, T. (2010). Citizens as sensors for natural hazards: A VGI integration workflow. *Geomatica, 64*(1), 41–60.

De Rubeis, V., Sbarra, P., Sorrentino, D., & Tosi, P. (2009). Web based macroseismic survey: Fast information exchange and elaboration of seismic intensity effects in Italy. In J. Landgren & S. Jul (Eds.), *Proceeding of the 6th international ISCRAM conference*. Gothenburg: University of Gothenburg.

Elwood, S., Goodchild, M. F., & Sui, D. Z. (2012). Researching volunteered geographic information: Spatial data, geographic research, and new social practice. *Annals of the Association of American Geographers, 102*(3), 571–590.

Farnham, S. D., Lahav, M., Raskino, D., Cheng, L., Ickman, S., & Laird-McConnell, T. (2012). So. cl: An interest network for informal learning. In *Proceedings of the sixth international AAAI conference on weblogs and social media* (pp. 90–97). Dublin: Association for the Advancement of Artificial Intelligence.

Goodchild, M. F. (2007). Citizens as sensors: The world of volunteered geography. *GeoJournal, 69*(4), 211–221.

Goodchild, M. F. (2009). NeoGeography and the nature of geographic expertise. *Journal of Location Based Services, 3*(2), 82–96.

Goodwin, K., Kenedy, G., & Vetere, F. (2010). Getting together out-of-class: Using technologies for informal interaction and learning. In *Proceedings ASCILITE* (pp. 387–392). Sydney: Ascilite.

Haklay, M. (2010). How good is volunteered geographical information? A comparative study of OpenStreetMap and Ordnance Survey datasets. *Environmental and Planning B: Planning and Design, 37*, 682–703.

Haklay, M. (2013). Citizen science and volunteered geographical information. Overview and typology of participation. In D. Z. Sui, S. Elwood, & M. F. Goodchild (Eds.), *Crowdsourcing geographic knowledge: Volunteered Geographic Information (VGI) in theory and practice* (pp. 105–122). Berlin: Springer.

Howland, J. L., Jonassen, D., & Marra, R. M. (2012). *Meaningful learning with technology*. Boston: Pearson.

Hughes, A. L., & Palen, L. (2009). Twitter adoption and use in mass convergence and emergency events. In J. Landgren & S. Jul (Eds.), *Proceeding of the 6th international ISCRAM conference*. Gothenburg: University of Gothenburg.

Ito, M. (2010). *Hanging out, messing around, and geeking out: Kids living and learning with new media*. Cambridge, MA: The MIT Press.

Jamieson, P. (2009). The serious matter of informal learning. *Planning for Higher Education, 37* (2), 18–25.

Jonassen, D. H., Howland, J. L., Moore, J. L., & Marra, R. M. (2003). *Learning to solve problems with technology: A constructivist perspective* (2nd ed.). Columbus: Merrill/Prentice Hall.

Jones, B. E., & Conceição, S. (2008). Can social networking tools foster informal learning? In *24th annual conference on distance teaching & learning*. Madison: University of Wisconsin.

Jordan, G. (2003). GIS for community participation in forestry user groups in Nepal: Putting people before the technology. In W. H. Craig, T. M. Harris, & D. Weiner (Eds.), *Community participation and geographic information systems* (pp. 232–245). London: Taylor and Francis.

Kingston, R. (2003). Web-based PPGIS in the United Kingdom. In W. H. Craig, T. M. Harris, & D. Weiner (Eds.), *Community participation and geographic information systems*. London: Taylor and Francis.

Knowles, M. S. (1980). *The modern practice of adult education. Andragogy versus pedagogy*. Cambridge: Prentice Hall.

Kwaku Kyem, P. A. (2003). Promoting local community participation in forest management through a PPGIS application in Southern Ghana. In W. H. Craig, T. M. Harris, & D. Weiner (Eds.), *Community participation and geographic information systems*. London: Taylor and Francis.

Lee, I., & Torpelund-Bruin, C. (2011). Geographic knowledge discovery from 2.0 technologies for advance collective intelligence. *Informatica, 35*, 453–461.

Lévy, P. (1999). *Collective intelligence: Mankind's emerging world in cyberspace*. New York: Perseus Books.

Muñiz-Solari, O. (2014). Geography education is a global responsibility. In *Comprehensive academic conference of the Chinese geography education national meeting*. Tianjin City, People's Republic of China.

Neis, P., Zielstra, D., & Zipf, A. (2013). Comparison of volunteered geographic information data contributions and community development for selected world regions. *Future Internet, 5*, 282–300.

O'Reilly, T. (2005). *What is web 2.0: Design patterns and business models for the next generation of software*. Blog by Tim O'Reilly retrieved from htttp://www.oreillynet.com/pub/a/oreilly/tim/news/2005/09/30/what-is-web-20.html

Pétursdóttir, S. D. D. (2011). *Technology enabled citizen participation in Nairobi slum upgrades*. Unpublished master thesis, Reykjavic University, Reykjavic.

Prandini, M., & Ramilli, M. (2012). Raising risk awareness on the adoption of Web 2.0 technologies in decision making processes. *Future Internet, 4*, 700–718.

Roger, A. (2004). Looking again at non-formal and informal education – Towards a new paradigm. In *The encyclopaedia of informal education*. www.infed.org/biblio/non_formal_paradigm.htm

Rogers, A. (2003). *What is the difference? A new critique of adult learning and teaching*. Leicester: NIACE.

Salisbury, D. S. (2011). *Transboundary mapping to reconcile conservation and development in the Amazon Borderlands: A case study from Peru-Acre, Brazil*. Poster presented at the Association of American Geographers Conference, Seattle, WA.

Schade, S., Luraschi, G., De Longueville, B., Coz, S., & Díaz, L. (2010). Citizens as sensors for crisis events: Sensor web enablement for volunteered geographic information. In *WebMGS 2010: 1st international workshop on pervasive web mapping, geoprocessing and services* (Vol. XXXVIII-4/W13, pp. 1–7). Como: ISPRS – International Society for Photogrammetry and Remote Sensing.

Surowiecki, J. (2004). *The wisdom of crowds: Why the many are smarter than the few and how collective wisdom shapes business, economies, societies, and nations*. New York: Doubleday.

Sylaiou, S., Basiouka, S., Patias, P., & Stylianidis, E. (2013). The volunteered geographic information in archeology. In *ISPRS annals of the photogrammetry, remote sensing and spatial information science, XXIV international CIPA symposium* (Vol. II-5/W1, pp. 301–306). Strasbourg: ISPRS.

Walker, D. H., Leitch, A. M., De Lai, R., Cottrell, A., Johnson, A. K. L., & Pullar, D. (2003). A community-based and collaborative GIS joint venture in rural Australia. In W. H. Craig, T. M. Harris, & D. Weiner (Eds.), *Community participation and geographic information systems*. London: Taylor and Francis.

Weiner, D., & Harris, T. M. (2003). Community-integrated GIS for land reform in South Africa. *URISA Journal, 15*(APA II), 62–73.

Werquin, P. (2010). *Recognising non-formal and informal learning: Outcomes, policies and practices*. Paris: Organization for Economic Co-operation and Development, OECD Publishing.

Part III
Teacher Education for Geospatial Technologies

Chapter 9
WebGIS in Education

Thomas R. Baker

Abstract Web-based GIS or WebGIS is powerful mapping and analytical functionality expressed within a web browser. Today, webGIS comes in many forms from consumer navigational maps to versatile location analytics tools that allow for user-directed analysis and content discovery. With nearly two decades of development history, webGIS tools are rapidly amassing strength, precision, and speed – in some cases, surpassing the capacities of basic desktop GIS applications. The inclusion of real-time data from sensor networks, social media, and the larger GIS community can extend WebGIS-based instruction to every learner. Because of the increased power and customization, webGIS can now better support learning standards-oriented content in the natural and social sciences. No longer is it necessary to teach learners how to use GIS before teaching the disciplinary content of interest.

In this article, the growing educative role for webGIS, both conceptually, technically, and practically will be explored, including the new affordances (e.g. collaboration, real-time data, distributed data, BYOD support, and interface customization) and constraints (e.g. bandwidth, privacy, and user management) provided by GIS in a cloud-based platform. Finally, considerations for preparing for new and pre-service teachers of webGIS, including pedagogical and technical considerations, will be discussed – providing a broad vision of the future of webGIS and how learners and educators can best utilize and prepare for that future.

Keywords WebGIS • Web mapping • GIS

9.1 Introduction

Web-based GIS or webGIS is powerful, geographic visualization and analytical functionality expressed within a web browser. In this chapter, "webGIS" will serve as an overarching term to mean web mapping, webGIS mapping, and Internet-based

---

T.R. Baker (✉)
Esri Inc, Redlands, CA, USA
e-mail: tbaker@esri.com

O. Muñiz Solari et al. (eds.), *Geospatial Technologies and Geography Education in a Changing World*, Advances in Geographical and Environmental Sciences,
DOI 10.1007/978-4-431-55519-3_9

GIS or IGIS – mapping and geographic analysis accessed using a mobile or desktop web browser. Today, webGIS comes in many forms from consumer navigational maps to versatile, location analytic tools that allow for user-directed analysis and content discovery. With nearly two-decades of development history, webGIS is rapidly amassing strength, precision, and speed – in some cases, surpassing the capacities of basic desktop GIS applications. The inclusion of real-time data from sensor networks, social media, and the larger GIS community can extend webGIS-based instruction to every learner. Because of the increased power and customization, webGIS can now better support learning standards-oriented content in the natural and social sciences. No longer is it necessary to teach learners how to use GIS before teaching the disciplinary content of interest.

This chapter explores the accumulated knowledge of webGIS and its successor technologies across education. The chapter discusses webGIS in improving student achievement, process and affective skills. Implications for best practices in teaching, learning, and curriculum development follow, based on research presented. Finally, and perhaps most importantly as it is least well documented, a discussion of the next evolution in webGIS, GIS as a platform.

9.2 What is WebGIS in Education?

WebGIS is typically viewed as an interactive map and underlying data that can be interrogated or analyzed by a user, requiring a modern web browser. WebGIS does basic things like zoom, pan, and identify but it also is capable or rich data analysis like creating density clusters or "hot spots" of activity or determining the geographic center of a particular set of data – all at the click of a button. Modern webGIS is frequently considered the frontend of a much larger geographic information system – where multiple Internet, mapping, and database servers can work in concert to provide maps and analysis to users. The modern webGIS can mash together data from multiple authoritative or social data sources, a so-called "mash-up".

WebGIS has been visible in K-12 education since the citizen science networks of the mid-1990s, such as GLOBE, KanCRN, Journey North, and others. These organizations and others led the development of collaborative webGIS in education, supporting interactive websites that allowed students to collect, analyze, and map their data (Baker 2005). At the same time, other large organizations were moving their paper-based atlases and maps into interactive, interrogative webGIS applications. For example, hugely popular in U.S. schools, the National Geographic launched its first online, interactive mapping tool, "The MapMachine" in 1999. More prominent in higher education, the Xerox PARC Map Viewer first started serving maps in 1993. These historical milestones in the development of webGIS represent a very wide range of applications, still largely describes the gamut today.

9.3 WebGIS and Learning: Research

The literature on the use of web mapping to support effective classroom instruction is clear: it can be an effective tool for enhancing student learning within appropriately designed learning experiences. (e.g. Bodzin and Anastasio 2006; Schultz et al. 2008; Milson 2011; Henry and Semple 2012). Adaptations of webGIS such as the story map, a blending of map, narrative, and digital media to tell a story, seem to show early promise an instructional tool (Strachan 2014). Moreover, when webGIS is used within an inductive, constructivist learning environment, like Project Based Learning, the results appear especially effective (Baker and White 2003; Milson and Earle 2008; King 2008; Huang 2011).

The positive learning effects of constructivism and webGIS (and geospatial technology [GST] generally) seem positive from late elementary through adult learners. Studies support web mapping as a tool for improving data analysis (Baker and White 2003; Bodzin and Anastasio 2006), and increasing cultural awareness (Milson and Earle 2008), to name a few skills. Researchers have also initially explored the positive effects of web mapping on improved "spatial thinking" (Manson et al 2013). Moreover, web mapping tools can decrease teacher-training time by using customized data and map interfaces (Baker 2005; Henry and Semple 2012; Huang 2011). For a more extensive review of the research on webGIS and GST generally, see the article, "Call for an Agenda and Center for GIS Education Research" (Baker et al. 2012). This discussion is further clarified with emerging research pathways in a recent, cross-disciplinary research agenda (Baker et al. 2015).

9.4 The Advantages of WebGIS in Education

WebGIS has many exciting features that stand it apart from previous geographic software tools for schools. Because *webGIS is delivered in a web browser*, the interfaces and data can be tailored to learner need, accounting for developmental level, instructional objectives, and even the hardware technology being used to access the maps or analysis. While most geographic desktop software is capable of streaming data to a user's computer, these applications don't necessarily use the web; rather they use the Internet for sending and receiving data and are therefore not a part of webGIS. The technical distinctions in the data protocols or transport mechanisms are largely arbitrary for instructional designers and educators. However, the critical capacity of a web browser to perform GIS data presentation and analysis cannot be understated.

The litany of reasons why desktop GIS in K-12 education never witnessed substantial adoption is long and in many cases, well deserved (Kerski 2003). The limitations presented by desktop software installation, data acquisition and storage, teacher training, school IT and administrative support, suitable hardware, and appropriately supportive curriculum were challenging, even for the most prepared schools.

Greater advocacy for “Science Technology Engineering and Math” (STEM fields), college and career readiness, computational thinking, and critical thinking continue to engender greater growth opportunities for webGIS. The more recent trend to, “Bring Your Own Device” (BYOD) or the case where schools are purchasing and deploying handheld tablets by the thousands, has further driven the adoption of webGIS. Tablets today are simply not intended to run applications like desktop GIS. It is relatively safe to say, the “golden age” of desktop GIS in schools is quietly passing. We are in a new age, with new tools and new expectations for learning. We are in the age of webGIS and “GIS as a platform”.

WebGIS presents many advantages for learning in formal and informal education, in schools and universities. While all technical and pedagogical advantages may not be present in all webGIS applications, many features are typically available in flagship webGIS tools (Fig. 9.1).

WebGIS has increased capacities for collaboration, analysis, storytelling, sharing, and interactivity. The analysis interfaces are simpler than ever, using wizards to guide novices through better decision-making. Today, the learner must know what a particular analysis is, why it is important, and how to interpret the results. There are relatively few confusing commands or obscure interfaces, requiring a GIS professional to navigate.

WebGIS and GIS platforms have ushered in the proliferation of *multi-scale data* for student inquiries. In a multi-scale data source, data in a single data service can change depending on the scale of the map. The data granularity, source, or

Fig. 9.1 A student collects field data with a GPS for later mapping in her webGIS (Worker 2014a)

Fig. 9.2 Field data collection with a webGIS running on a device (Worker 2014b)

classification can change as a student zooms in or out of the data set. Multi-scale data makes inquiries at many more geographic scales possible. It also provides cartographically proper data at any level of investigation. Multi-scale data allows an instructor the leeway to ask either, “What is the dominant religion across Africa?” or “What is the dominant religion within Blantyre, Malawi?” – all from the same data service.

WebGIS increases support for user-contributed data (e.g. citizen science, crowdsourcing, or volunteered geographic information). WebGIS enables not only display of the data but also the analysis of that data, as it arrives at the server or at a developer-specified interval. Learners now use mobile apps on smartphones to collect data to the webGIS, viewable by other students and the teacher back on smartphones, tablets, or desktop computers (Fig. 9.2).

For example, in informal education, youth work with scientists from the U.S. Fish and Wildlife Service to collect point locations of invasive species across the Tijuana Slough, near San Diego, California. After mapping the occurrences of the invasive species with a tablet device, the data are transmitted back to ArcGIS Online for further analysis in a mobile lab. In time, U.S. Fish and Wildlife will return to the slough using the maps and analysis to target their invasive species control methods. Similar community-focused projects (e.g. biodiversity monitoring, debris inventories, etc.) in other countries can be seen through organizations like GeoPorter (http://geoporter.net).

WebGIS allows for customized learner experiences, as the webGIS interface can be designed specifically for the student’s task. This presents profound possibilities for teaching and learning. “The dynamic display and connection of spatial information enhances learning efficiency. Map scaling and partitioning reduces the cognitive load imposed on the learner and just-in-time association of maps with non-spatial facts in instruction media facilitates comprehension of the subject matter” (Huang 2011, p. 164).

WebGIS can increase mobile device value, using on-board GPS to access web-based maps and data. Learners can carry their GIS on their phone – or soon perhaps, even a watch or glasses (called “wearable tech”). The prevalence of location-based technologies (including GPS), turn many modern digital devices into a field data collection device. For example, see the Esri Collector, EduLoc, Ushahidi, or a host of other geo-enabled mobile apps for data collection. Mobile webGIS is increasingly serving as a tool for making decisions, as a mobile data dashboard. While mobile devices can frequently access or contribute map-based data via a native application, mobile devices can also access webGIS applications, designed for a mobile browser. These mobile webGIS applications don’t require software installation, software fees, or mobile OS compatibility.

WebGIS reduces the need for handling large, complicated, or changing files, including remotely sensed imagery. For example, why download terabytes of imagery of your state or county when students can stream to their computers or devices the compressed data they need, when it’s needed? In cases of tablets, streaming data as it is needed is the standard today, as tablets have comparatively little disk space for file storage.

WebGIS can eliminate the need for complex desktop operating systems and steep hardware requirements. WebGIS runs equally well in Windows, Macintosh, Linux, or common mobile operating systems like Android, iOS, and Windows Mobile. Stringent memory, hard drive, and video card requirements are a thing of the past. Moreover, the need for IT staff on campus is diminished. And perhaps most importantly, no software installations or upgrades are necessary because when the webGIS is updated on the server by a GIS developer, changes rollout invisibly to students and educators.

Educational users can access and use webGIS from home, school, or nearly anywhere. This new ease of access complements a recent trend in instructional design, where educators put direct instruction online (often in the form of pre-recorded videos) for evening viewing by students. In this “flipped” classroom, students do traditional homework (or activities) during the regular class meeting. The flexibility of webGIS to be used anywhere, making it an ideal candidate for supporting a wide range of instructional approaches.

For example, an educator can provide minimal direction to students as they create a community “food desert” map, consisting of population density, grocery stores and farmers markets, a mobility index – all layered atop a remotely sensed image (Fig. 9.3).

Researchers and authors have declared that webGIS is one of the key reasons that GIS is gaining ground in the US and internationally (Milson and Kerski 2012). Understanding webGIS adoption across education might be summarized with a few key milestones. Of the 60 University Consortium for GIS (UCGIS) member

Fig. 9.3 Students use a webGIS on a desktop computer (Worker 2014c)

institutions, creating and sharing webGIS maps is easily described as commonplace. Of the over 7000 institutions of higher education globally that use Esri products, the majority of these create and use web maps for instruction. In the US, it is anticipated that by 2017, well over 25,000 schools will use webGIS in instruction – at a variety of instructional levels and in many disciplines. Globally, the primary and secondary markets are more difficult to quantify, although schools in Canada, the UK, and Europe (especially Germany, Norway, and France) are increasingly creating webGIS maps, especially in geography and Earth or environmental science. A 2012 international collection of desktop and webGIS adoption stories bears similar conclusions, noting national initiatives or interesting adoption models in a variety of countries including Australia, Taiwan, Turkey, Rwanda, Dominican Republic, and Germany (Milson et al. 2012).

WebGIS supports improved real-time data (e.g. crowd-sourced data or data from sensor networks). WebGIS has been used in citizen science networks since at least the mid 1990s. Today that technology has grown up, capable of handling increased volumes of users more complicated data, and real-time analysis. For example, data from social media can be used as crowd sourced information to inform the broader public about natural hazards, social movements, or the whereabouts of the latest teen idol.

Finally, webGIS data and tools are fast and engaging for learners. New, wizard-driven analysis tools in webGIS (e.g. ArcGIS Online for Organizations) allows students to create hot spot analyses, "geoenrich" existing data, overlay analysis, or derive best locations – all with a few clicks. Analysis is much more than distance measurements or dropping a marker on a map! Analysis produces new information about patterns and trends in data, when guided by an informed user.

9.5 Challenges to Best Practices

WebGIS in education may still present some challenges, including bandwidth, privacy, and student-user account management. These challenges will vary by school, local, regional or national policies and technology infrastructures. And while these challenges will no doubt improve, they will take time and effort to improve.

Broadband access may be the largest challenge for classrooms worldwide, when it comes to implementing web-based GIS. By 2009, 93 % of U.S. classroom computers were reported to have Internet access (NCES). However, the quality of the access is often insufficient for multiple computers to access the maps, data, or analytics provided by a webGIS. Worldwide, school networks are extremely varied and performance localized. For example, resolving this patchwork of educational bandwidth in the U.S. alone is a top priority of the Obama Whitehouse and the connectEd program. "The ConnectED Initiative will, within 5 years, connect 99 % of America's students to the digital age through next-generation broadband and high-speed wireless in their schools and libraries" (The United States Whitehouse 2014). While other developed nations have similar programs in place, in many cases developing nations struggle to provide the necessary computer hardware or Internet access in schools.

Student privacy on the Internet is paramount. In many older educational networks, students contributed data to a web-form or webGIS, often lacking the privacy and security demanded under U.S. Federal laws such as the Children's Online Privacy Protection Act (COPPA) and Family Educational Rights and Privacy Act (FERPA). Today, newer, more powerful webGIS systems can create a "walled garden" around a school's geospatial data and analysis – allowing only students, parents, and staff to view or interrogate the details of geographic data. These same modern webGIS systems can also allow for multiple levels of permission or security depending upon who the user might be – allowing for different users to have access to varying levels of information. The challenge for most schools implementing role-based permissions is the need for user management, including account creation for hundreds of students and dozens of faculty. While the pragmatics of user management are improving with each passing month, it remains a substantial challenge, especially in larger public school systems and universities. Through the use of "walled gardens" and user account management, some modern webGIS tools can protect minors while still allowing for the full range of geospatial tools and learning needed.

9.6 The Future: From WebGIS to the Platform

"GIS as a platform" is the evolution of GIS technology. The platform relies on the "cloud" and any number of users, devices (e.g. tablets, phones, wearable tech), or software applications creating, interrogating, and visualizing location-based information. The GIS "cloud" uses multiple arrays of redundant and powerful Internet servers acting in concert. The GIS cloud runs software capable of doing many of the things traditional desktop GIS can do, as well manage users, support collaboration, enforce security, and more. The result is that a learner can record a piece of information while in the community (e.g. report graffiti) and local community enforcement officers can see the data and its analysis on a map in a crime lab or in a police car. Moreover, social media channels can be federated to produce real-time listings of "geo-events" (events or situations in a community tagged with location) can be placed on the map – all to paint a better picture of what's going on, in real-time.

GIS as a platform will be increasingly advantaged by its capacity to collect, analyze, and visualize massive amounts of data from automated sensors to crowdsourcing networks. GIS as a platform today allows novices to drape high-resolution imagery across high-resolution imagery. Generate three and four-dimensional map-based visualizations for making better decisions in nearly any field of study.

Are teachers and students ready to capitalize on WebGIS? A way to consider teacher readiness is through the lens of TPCK – Technology-Pedagogy-Content-Knowledge (Mishra and Koehler 2006; Koehler and Mishra 2009). In short, readiness is dependent on teachers' knowledge of pedagogy, content, and technology – but especially critical is mastery of the interaction of these elements within the context of GST (Bodzin et al. 2012; Doering et al 2014; MaKinster and Trautmann 2014). In effect, if educators do not have mastery of the interaction effects of these three broad factors, their GST-based instruction may not succeed. Thankfully, in the era of webGIS, the technology component of TPCK can be less burdensome and the interaction of the TPCK factors much more efficient to master. However, in this case, the "technology" includes GIScience, as in so far it is relevant to implementing webGIS. This could include understanding basic database concepts, cartography and statistics. For example, if students do not understand how and when to use normalized data, they will make an equally poor map in either desktop or webGIS. Perhaps in the near future, GIS systems will become intelligent enough to provide "just in time" instruction to avoid these and other cartographic blunders.

9.7 Closing

WebGIS tools are easier, faster, and more powerful than ever before; they are at the fingertips of educators and students alike in many places around the world. Educational research supporting the use of WebGIS is promising. From enhancing student learning outcomes and affect to reducing teacher preparation time and technical challenges, the tools and methods have proven their worth. The capacity for webGIS to support educational priorities like STEM and career readiness, instructional practices like Project Based Learning and constructivism, and overcome previous technical issues like data scale, size, and complexity all add to the advancement of webGIS in education. While a few challenges remain, largely due to bandwidth, these will continue to improve as that bandwidth becomes more ubiquitous, stable and faster. The influences of webGIS will, if it hasn't already, change the landscape of geographic and science education in schools and universities around the world.

References

Baker, T. R. (2005). Internet-based mapping to support K12 education. *The Professional Geographer, 57*(1), 44–50.

Baker, T. R., & White, S. H. (2003). The effects of geographic information system (GIS) technologies on students' attitudes, self-efficacy, and achievement in middle school science classrooms. *Journal of Geography, 102*(6), 243–254.

Baker, T. R., Kerski, J. K., Huynh, N. T., Viehrig, K., & Bednarz, S. W. (2012). Call for an agenda and center for GIS education research. *Research of International Geographical Education Online, 2*(3), 254–288.

Baker, T. R., Battersby, S., Bednarz, S. W., Bodzin, A. M., Kolvoord, R., Moore, S., Sinton, D., & Uttal, D. (2015). A research agenda for geospatial technologies and learning. *Journal of Geography, 114*(3), 118–130.

Bodzin, A. M., & Anastasio, D. (2006). Using web-based GIS for earth and environmental systems education. *Journal of Geoscience Education, 54*(3), 295–300.

Bodzin, A., Peffer, T., & Kulo, V. (2012). The efficacy of educative curriculum materials to support geospatial science pedagogical content knowledge. *Journal of Technology and Teacher Education, 20*(4), 361–386.

Doering, A., Koseoglug, S., Scharberg, C., Henrickson, J., & Lanegran, D. (2014). Technology integration in K-12 geography education using TPACK as a conceptual model. *Journal of Geography, 113*(6), 223–237.

Henry, P., & Semple, H. (2012). Integrating online GIS into the K–12 curricula: Lessons from the development of a collaborative GIS in Michigan. *The Journal of Geography, 111*(1), 3–14.

Huang, K. H. (2011). A GIS-Interface web site: Exploratory learning for geography curriculum. *The Journal of Geography, 110*(4), 158–165.

Kerski, J. J. (2003). The implementation and effectiveness of GIS in secondary education. *Journal of Geography, 102*(3), 128–137.

King, E. (2008). Can PBL-GIS work online? *The Journal of Geography, 107*(2), 43–51.

Koehler, M., & Mishra, P. (2009). What is technological pedagogical content knowledge (TPACK)? *Contemporary Issues in Technology and Teacher Education, 9*(1), 60–70.

MaKinster, J., & Trautmann, N. (2014). Understanding the use of geospatial technologies to teach science: TPACK as a lens for effective teaching. In A. Makinster, A. Trautman, & A. Barnett (Eds.), *Teaching science and investigating environmental issues with geospatial technology: Designing effective professional development for teachers*. New York: Springer.

Manson, S., Shannon, J., Eria, S., Kne, L., Dyke, K., Nelson, S., Batraa, L., Bonsal, D., Kernik, M., Immich, J., & Matson, L. (2013). Resource needs and pedagogical value of web mapping for spatial thinking. *The Journal of Geography, 113*(1), 1–11.

Milson, A. J. (2011). The cultivation of spatial-civic decision-making through WebGIS. In T. Jekel, A. Koller, K. Donert, & R. Vogler (Eds.), *Learning with geoInformation: Implementing digital earth in education* (pp. 12–18). Berlin: Wichmann.

Milson, A., & Earle, B. (2008). Internet-based GIS in an inductive learning environment: A case study of ninth-grade geography students. *The Journal of Geography, 106*(6), 227–237.

Milson, A., & Kerski, J. (2012). Around the world with geospatial technologies. *Social Education, 76*(2), 105–108.

Milson, A., Demirci, A., & Kerski, J. J. (Eds.). (2012). *International perspectives on teaching and learning with GIS in secondary schools*. Dordrecht: Springer.

Mishra, P., & Koehler, M. J. (2006). Technological pedagogical content knowledge: A framework for teacher knowledge. *Teachers College Record, 108*(6), 1017–1054.

National Center for Educational Statistics (NCES). (2009). *Fast facts*. U.S. Department of Education. http://nces.ed.gov/fastfacts/display.asp?id=46. Accessed 1 Mar 2014.

Schultz, R., Kerski, J., & Patterson, T. (2008). The use of virtual globes as a spatial teaching tool with suggestions for metadata standards. *Journal of Geography, 107*(1), 27–34.

Strachan, C. (2014). *Teacher's perceptions of Esri story maps as effective teaching tools* Master's thesis. Retrieved from the University of South Carolina.

The United States Whitehouse. (2014). *The Whitehouse Blog*. The United States Whitehouse. http://www.whitehouse.gov/blog/2013/06/06/what-connected. Accessed 1 Mar 2014.

Worker, E. (2014a). *A student collects field data with a GPS for later mapping in her webGIS* [image]. Provided by E. Worker.

Worker, E. (2014b). *Field data collection with a webGIS running on a device* [image]. Provided by E. Worker.

Worker, E. (2014c). *Students use a webGIS on a desktop computer* [image]. Provided by E. Worker.

Chapter 10
Teaching GIS and Other Geospatial Technologies to In-Service Teachers

Jung Eun Hong

Abstract Since the early 1990s, geospatial technologies (GSTs) have been reported as effective and useful instructional tools in K–12 education to improve students' learning and enhance their critical thinking, spatial thinking, and problem solving skills. However, the number of teachers who are implementing those technologies in their classrooms is still pretty low. Among various identified barriers, the major ones are teachers' lack of background and lack of time to learn, practice, and develop lesson plans using GSTs. In order to solve these barriers, many researchers and professionals in different countries have been providing various kinds of training for in-service teachers. In this chapter, six models of in-service teacher training of GSTs—project-based learning, community partnership, iterative training, minimal Geographic Information System, snowball dispersion, and online training—are introduced and reviewed with exemplary case studies conducted in several countries. The future direction of the effective and useful in-service teacher training is also discussed.

Keywords Geospatial technology • Teacher training • Instructional technology

10.1 Introduction

Since the early 1990, many scholars have demonstrated the effectiveness and usefulness of geospatial technologies (GSTs) as instructional tools in K–12 education. However, their adoption rate in the K–12 classroom remains fairly low (Kerski 2003). Among identified barriers, the most consistently indicated problem is a dearth of teachers' knowledge, experiences, and background of GSTs (Bednarz 2004; Demirci et al. 2013; Kerski 2003; Meyer et al. 1999). The main reason is that not many in-service teachers have had an opportunity to learn about GSTs in their pre-service teacher education (Kerski 2003). Only a few universities offer a social studies methodology course with GSTs for their students. As a result, a majority of

J.E. Hong (✉)
Department of Geosciences, University of West Georgia, Carrollton, GA 30118, USA
e-mail: jhong@westga.edu

O. Muñiz Solari et al. (eds.), *Geospatial Technologies and Geography Education in a Changing World*, Advances in Geographical and Environmental Sciences,
DOI 10.1007/978-4-431-55519-3_10

pre-service teachers become in-service teachers without acquiring an educational background in GSTs.

Therefore, offering in-service teacher training has been suggested as an alternative solution to increase the adoption rate of GSTs in the K–12 classroom. Many organizations and professionals have been providing various GST trainings for in-service teachers by adopting several training models. In this chapter, six in-service teacher training models—project-based learning, community partnership, iterative training, minimal Geographic Information System (GIS), snowball dispersion, and online training—are introduced with exemplary case studies conducted in several countries. At the end of this chapter, the future direction of GST training for in-service teachers is discussed.

10.2 Six Models for In-Service Teacher training of GSTs

10.2.1 Project-Based Learning

Project-based learning (PBL) is a useful and effective model for student learning. The PBL model helps students become active learners. Through a PBL activity, students are able to (1) identify a problem, (2) collect and manage related data to solve the problem, (3) analyze and visualize the data, (4) make a decision based on the results they found, and (5) share their findings with other people (Blumenfeld et al. 1991). PBL also supports the effective learning and understanding of content material; students gain in-depth knowledge of the content and enhance their learning by solving real issues (Blumenfeld et al. 1991; Krajcik et al. 1994). In addition, a PBL activity also allows students to understand and engage with their community better when they choose an issue in their local community (Demirci et al. 2013; Meyer et al. 1999). Students also learn various social and scientific methods of collecting, analyzing, and presenting data, such as interviews, surveys, and maps (Demirci et al. 2011). Finally, students can improve responsibility, critical-thinking, problem-solving, and collaboration skills through this process (Keiper 1999; Kerski 2008).

Since PBL has been proven as an effective model in K–12 education, many researchers in different countries have adopted the PBL model for implementing GSTs in the K–12 classroom. For example, students in Turkey worked on a GIS project to learn about plates, earthquakes, and volcanoes (Demirci 2008). In Austria, students used GSTs for a climate change project (Jekel et al. 2012). As part of a social studies classroom project, Japanese students made a map using GIS to request more streetlights in their neighborhood (Ida and Yuda 2012).

According to Wilder et al. (2003), PBL is effective not only for students but also for teachers because it provides active learning environments. Rather than using pre-made data and lessons using GSTs, teachers can choose their own topics and collect related data in PBL. Additionally, teachers learn how to manipulate data to make it usable in GSTs. With the PBL training model, teachers also learn a successful role as a PBL facilitator (Blumenfeld et al. 1991). For example, instead of answering students' questions directly, teachers can lead students in the direction of self-learning. Last but not least, with the PBL teacher training model, teachers can practice integrating GSTs into their lessons effectively (Wilder et al. 2003). In other words, instead of learning how to teach *about* GSTs, teachers learn how to teach *with* GSTs (Sui 1995), which is recommended for GSTs in K–12 education.

Therefore, several scholars have used PBL for GST in-service teacher training and encouraged teachers to implement GST-based PBL in their classrooms. One of the GST teacher training examples based on PBL is "*Science Beyond the Classroom: Integrating GIT*[1] *with Project-Based Science*" in New Mexico, U.S. (Wilder et al. 2003). A total 27 secondary school science teachers in New Mexico participated in a two-year GST professional development program based on PBL. Teachers attended a 15-day main session during each summer and a 2-day follow-up session each fall for two years (2001–2002). In this training, participating teachers learned GSTs while working on a project about ultraviolet (UV) issues in New Mexico using GIS, Global Positioning System (GPS), and Remote Sensing (RS).

10.2.2 Community Partnership

Community partnership is a model based on collaborative partnership between a school district and an institution of higher education. This model was originally developed for pre-service teacher education (Pierson and McNeil 2000), but can also serve as an effective model for in-service teacher training in a new technology. In this model, in-service teachers supported by school administrators collaborate with pre-service teachers supported by college faculty members to build successful learning environments. The ultimate goal of the community partnership model is to enhance students' learning. In-service and pre-service teachers cooperate to identify needs in the classroom and to find solutions to meet the needs.

Both in-service and pre-service teachers can benefit from community partnership. In-service teachers can learn up-to-date technologies easily with direct support from relatively technology-experienced pre-service teachers to improve their existing curriculum. In the case of pre-service teachers, they can obtain real experience preparing lessons with the advice of in-service teachers. Pre-service teachers can also apply knowledge and skills that they learned at a college, such as lesson planning and technology integration, in the K–12 classroom. Therefore, collaboration between these two groups can improve education on both sides (Yu et al. 2011).

[1] GIT stands for geographic information technologies.

In 2000–2001, Alibrandi and Palmer-Moloney (2001) adopted this model for implementing GIS in the existing middle school social studies curriculum. Pre-service teachers, who took the Social Studies Teaching Methods course at Hartwick College in New York, U.S., partnered with in-service social studies teachers at Oneonta Middle School. First, pre-service teachers learned GIS, and then they met in-service teachers to choose a specific unit to be developed with GIS. Pre-service teachers collected and provided GIS resources for in-service teachers, and both groups worked together to develop GIS-based lessons. The community partnership model was successfully implemented in terms of bridging a gap and building cooperative networks between in-service and pre-service teacher education.

10.2.3 Iterative Training

The iterative training model supports the development of successful in-service teacher training through continuous modification based on feedback from teachers (Buss et al. 2002). Multiple iterations of teacher training occur in this model. Participating teachers' comments and suggestions are used to revise and strengthen the next iteration. This continuous revision persists until the last iteration to create an effective teacher training.

Using the iterative training model, a 5-year GIS in-service teacher training, *Earth System Science Internet Project (ESSIP)* was developed from 1996 to 2001 in Wyoming, U.S. (Buss et al. 2002). Fifteen secondary teachers attended the first workshop in 1996. After the workshop, teachers tested the lessons that they learned at the workshop in their classrooms. Then the project directors collected feedback from teachers. For successful classroom implementation, individual teachers received technical support from the project facilitators. A facilitator visited each teacher's classroom or answered a question via an e-mail or a phone call. After the classroom implementation, teachers took an interview to offer comments and suggestions for modifying the future workshop content. These teachers' comments were used as the key resources for the second iteration. A total of 21 teachers participated in the second workshop in 1997. Those teachers were also asked to test the GIS-based lessons that they developed during the workshop in their classrooms. Their feedback was used to revise the next workshop for the third iteration. A series of workshops were offered as the third iteration from 1999 to 2001, and 59 teachers were able to implement GIS-based lessons in the classroom successfully.

This model focuses on actively collecting feedback from teachers regarding the training in each stage, and then applying teachers' feedback for the next training in order to develop a better teacher training than the previous one. This model offers a long-term perspective and plan for GIS teacher training to develop an effective and useful training for in-service teachers.

10.2.4 Minimal GIS

The minimal GIS model emphasizes reducing the learning curve of GIS while improving spatial thinking skills (Marsh et al. 2007). This model helps focus on the students' spatial thinking concepts using an easy-to-learn and easy-to-use customized GIS application rather than a professional GIS software package, the one GIS experts use for their work. One of the main reasons for using GIS in K–12 education is to improve students' spatial thinking skills. However, the challenge of using difficult GIS software can inhibit students from building these spatial thinking skills.

Besides, the use of professional GIS software for in-service teacher training has been one of the barriers to classroom implementation of GIS (Doering and Veletsianos 2008; Kerski 2003; Liu and Zhu 2008; Meyer et al. 1999). Many in-service teacher trainings have used a professional GIS software package. Because the majority of in-service teachers do not have GIS experience and background, they have had difficulties mastering the professional version of GIS software.

For this reason, several researchers have developed customized GIS applications for K–12 education instead of using professional GIS software. One example of these applications is *World Explorer*, developed for secondary geography students in Singapore (Liu and Zhu 2008). Liu and Zhu designed *World Explorer* based on secondary geography content in Singapore and provided related maps and statistical data to support the content. Using resources provided in *World Explorer*, teachers were able to develop GIS-supported classroom activities easily. The user interface of *World Explorer* was designed for secondary-level students; only necessary tools and functions were available on the menu. Therefore, students could focus on learning content without spending too much time learning the application itself.

Henry and Semple (2012) also developed an online GIS tool, *$H_2OMapper$*, based on the minimal GIS model, and used it for a two-year teacher training project in Michigan, U.S. starting in 2008. *$H_2OMapper$* was a tool for teaching and learning about watersheds in Michigan, including water quality and land use, targeted at the 7th to 9th grades. *$H_2OMapper$* did not provide a full package of GIS analysis functionality. Instead, it only offered basic functions useful for learning about watersheds. The goal of *$H_2OMapper$* was to provide a user-friendly application and interface teachers and students could easily learn and use. With the customized, user-friendly GIS application, both teachers and students could reduce the time required to learn about GIS. Therefore, teachers could focus on implementing GIS into their curriculum, and students could develop their spatial thinking skills.

10.2.5 Snowball Dispersion Model

In the snowball dispersion model, a small group of teachers attend a teacher training and then train other teachers themselves. The goal of this model is to expand the skills and knowledge of a large group of teachers in a cost-effective way. In 2008, the Center for Geographic Information Systems and Remote Sensing of the National University of Rwanda (CGIS-NUR) adopted the snowball dispersion model and started to train 30 teachers in 10 secondary schools in Rwanda as the first phase (Forster and Mutsindashyaka 2008). After attending the training, teachers were able to introduce GSTs to approximately 500 students. During a second phase, CGIS-NUR led those 30 teachers in pilot schools to train other teachers in different schools and to support each other in increasing classroom implementation of GSTs in Rwanda. Using this training model, the research team expected to train teachers in approximately 600 secondary schools in Rwanda by 2011.

The snowball dispersion model is well-connected to the concept of teacher-consultants that Binko (1989) suggested. Teacher-consultants are teachers who train other teachers by offering in-service workshops. According to Binko (1989), the teacher-consultants approach is an effective way to spread knowledge and new skills to large group of teachers and their students. Similarly, the learning cluster model (LCM) has been proposed by Brysch and Boehm (2014). LCM encourages participating teachers to become teacher leaders who can train their colleagues. In reality, not all in-service teachers can receive a teacher training due to various reasons, such as lack of funding or personal issues. Like the snowball dispersion model, LCM allows more and more in-service teachers can learn about GSTs without investing much cost and time. Teachers can also form a community to share resources and support each other in developing and implementing GST-based lessons and activities.

10.2.6 Online Training

The online training model has been suggested as an alternative training model to address several concerns raised by the traditional face-to-face format, such as training costs, time and space restriction, and different learning paces among participating teachers (Borko et al. 2009; Frazier and Boehm 2012; Jung 2005). Therefore, many researchers and practitioners have adopted the online format to train various subjects and skills to in-service teachers. Previous studies have found that an online training was more or at least equally effective compared to a traditional face-to-face training in terms of learning performance, teaching knowledge, and teachers' satisfaction (Jung 2005; Liu et al. 2007; Pomales-García and Liu 2006; Pryor and Bitter 2008).

However, the online format has been not widely adopted for GST teacher training. To develop a successful online GST in-service teacher training, some important issues must be considered (Hong 2014). First, online training materials should be designed to be easy to follow and understand, since no direct personnel support is available. Second, teachers' different levels of computing ability need to be considered. Third, a way to communicate between trainers and trainees, such as online discussion board should be available in case teachers have a question to ask.

In the online teacher training, a web-based seminar (webinar) format is often used. One of the representative webinar programs for GSTs in education is developed by the National Council for Geographic Education (NCGE). NCGE provides 10–20 webinars per academic year for K–12 geography teachers. These include webinars on GSTs, as well as others addressing general geographical content and pedagogical skills. GST related webinars have mostly been presented by the Esri Education Team, but other GST professionals have also presented webinars. With a live webinar, audiences are able to ask questions directly to a presenter during the webinar presentation. For those audiences who missed a live webinar, archived webinars are also available on the NCGE website (http://www.ncge.org/webinar-archives-list) (National Council for Geographic Education 2014). The webinar training provides an interactive online learning environment for teachers, who are interested in learning GSTs but cannot attend a face-to-face workshop.

A tutorial type of online teacher training has also been developed. One example is *GIS for Social Studies*, developed for mainly middle school social studies teachers (Hong 2014). With consideration of the above online training issues, Hong adopted the user-centered design (UCD) approach to create user-centered and user-friendly tutorials so that teachers can understand and follow the tutorials alone. To do this, Hong interviewed secondary social studies teachers in Colorado, U.S. to analyze their needs and concerns. The first-draft version of the tutorials was developed based on the results of teacher-needs analysis. Then Hong met teachers again to collect feedback regarding the tutorials, and the tutorials were revised in accordance with this feedback to develop teacher-friendly GIS training tutorials. As a result, the majority of teachers who evaluated the tutorials found them easy to follow, and responded that they were able to complete the tutorials by themselves.

10.3 Conclusion and Future Direction

This chapter only introduces a few case studies of GST in-service teacher training based on six training models. Because each model has its own advantages and limitations, there is no single ideal training model for GSTs. Different training models may be particularly well-suited to different circumstances. We also need to take into consideration how we can combine the above six models to design a successful teacher training in the future. For example, rather than a sporadic, one-time training, school districts could offer a long-term, ongoing training for at least one academic year to communicate with teachers and support them in implementing GSTs in their classrooms.

Maintaining close partnership between school districts and higher education institutions is important for improving students' learning and the quality of teacher education in general. In addition, strong partnership with industries is also recommended. All three agencies may benefit from the partnership: school districts and higher education institutions may receive continuous technical support from GST industries, while the industries can develop effective applications for K–12 students with feedback from teachers. Industries' excessive involvement may cause some issues, such as expansion of a single application in K–12 education and hindrance of creative applications and materials developed by individual researchers. However, close cooperation among three organizations may help design effective teacher training and develop useful resources.

Opportunities to learn GSTs for in-service teachers still differ from country to country (Kerski et al. 2013). Teachers in countries where GSTs have been widely introduced and incorporated into the curriculum, such as China, Denmark, Japan, Taiwan, Turkey, and the U.S., have access to various training opportunities and resources. However, there is insufficient support for teachers in many developing and some developed countries due to lack of technology equipment for K–12 education and lack of interest in GSTs as instructional tools.

In order to encourage teachers to use GSTs in their classrooms, offering professional development opportunities is vital. Moreover, we need to provide learning-friendly environments for teachers in the future so that they can learn and use GSTs actively. A one-stop resource center may help teachers find materials that they need easily. So far, many individual projects have developed and provided valuable resources for teachers, but they are not in one place rather scattered here and there. Therefore, teachers may have a hard time searching for specific materials that they require. If materials in a one-stop resource center were maintained well and continuously updated, teachers would find them useful. An online community for teachers could also be a great resource. This online community could be a place to share materials with each other, such as classroom activities, pedagogy, and assessments, and to discuss better integration of GSTs in K–12 education. Various learning opportunities for teachers may lead to an increase in the adoption rate of GSTs in K–12 education.

References

Alibrandi, M., & Palmer-Moloney, J. (2001). Making a place for technology in teacher education with geographic information systems (GIS). *Contemporary Issues in Technology and Teacher Education, 1*(4), 483–500.

Bednarz, S. W. (2004). Geographic information systems: A tool to support geography and environmental education? *GeoJournal, 60*(2), 191–199.

Binko, J. B. (1989). *Spreading the word about geography: A guide for teacher-consultants in geography education*. Washington, D.C.: National Geographic Society.

Blumenfeld, P. C., Soloway, E., Marx, R. W., Krajcik, J. S., Guzdial, M., & Palincsar, A. (1991). Motivating project-based learning: Sustaining the doing, supporting the learning. *Educational Psychologist, 26*(3–4), 369–398.

Borko, H., Whitcomb, J., & Liston, D. (2009). Wicked problems and other thoughts on issues of technology and teacher learning. *Journal of Teacher Education, 60*(1), 3–7.

Brysch, C. P., & Boehm, R. G. (2014). Online professional development in geography: The learning cluster method and teacher leadership. *European Association of Geographers, 5*(1), 62-69.

Buss, A., McCLlurg, P., & Dambekalns, L. (2002). An investigation of GIS workshop experiences for successful classroom implementation. In *Proceedings of the ESRI education GIS conference*. San Diego: ESRI.

Demirci, A. (2008). Evaluating the implementation and effectiveness of GIS-based application in secondary school geography lessons. *American Journal of Applied Sciences, 5*(3), 169–178.

Demirci, A., Karaburun, A., Ünlü, M., & Özey, R. (2011). Using GIS-based projects in learning: Students help disabled pedestrians in their school district. *European Journal of Geography, 2* (2), 48–61.

Demirci, A., Karaburun, A., & Ünlü, M. (2013). Implementation and effectiveness of GIS-based projects in secondary schools. *Journal of Geography, 112*(5), 214–228.

Doering, A., & Veletsianos, G. (2008). An investigation of the use of real-time, authentic geospatial data in the K–12 classroom. *Journal of Geography, 106*(6), 217–225.

Forster, M., & Mutsindashyaka, T. (2008). Experiences from Rwandan secondary schools using GIS. In *Proceedings of the ESRI education user conference*. San Diego: ESRI.

Frazier, C. A., & Boehm, R. G. (2012). Using technology for geography teacher education: Web-based professional development. *Review of International Geographical Education Online, 2*(1), 78–94.

Henry, P., & Semple, H. (2012). Integrating online GIS into the K–12 curricula: Lessons from the development of a collaborative GIS in Michigan. *Journal of Geography, 111*(1), 3–14.

Hong, J. E. (2014). Promoting teacher adoption of GIS using teacher-centered and teacher-friendly design. *Journal of Geography*. doi:10.1080/00221341.2013.872171.

Ida, Y., & Yuda, M. (2012). Japan: GIS-enabled field research and a cellular phone GIS application in secondary schools. In A. J. Milson, A. Demirci, & J. J. Kerski (Eds.), *International perspectives on teaching and learning with GIS in secondary schools* (pp. 141–149). New York: Springer.

Jekel, T., Koller, A., & Strobl, J. (2012). Austria: Links between research institutions and secondary schools for geoinformation research and practice. In A. J. Milson, A. Demirci, & J. J. Kerski (Eds.), *International perspectives on teaching and learning with GIS in secondary schools* (pp. 27–36). New York: Springer.

Jung, I. (2005). Cost-effectiveness of online teacher training. *Open Learning: The Journal of Open, Distance and e-Learning, 20*(2), 131–146.

Keiper, T. A. (1999). GIS for elementary students: An inquiry into a new approach to learning geography. *Journal of Geography, 98*(2), 47–59.

Kerski, J. J. (2003). The implementation and effectiveness of geographic information systems technology and methods in secondary education. *Journal of Geography, 102*(3), 128–137.

Kerski, J. J. (2008). The role of GIS in digital earth education. *International Journal of Digital Earth, 1*(4), 326–346.

Kerski, J. J., Demirci, A., & Milson, A. J. (2013). The global landscape of GIS in secondary education. *Journal of Geography, 112*(6), 232–247.

Krajcik, J. S., Blumenfeld, P. C., Marx, R. W., & Soloway, E. (1994). A collaborative model for helping middle grade science teachers learn project-based instruction. *The Elementary School Journal, 94*(5), 483–497.

Liu, S., & Zhu, X. (2008). Designing a structured and interactive learning environment based on GIS for secondary geography education. *Journal of Geography, 107*(1), 12–19.

Liu, C. H., Chiang, T. C., & Huang, Y. M. (2007). Assessment of effectiveness of web-based training on demand. *Interactive Learning Environments, 15*(3), 217–235.

Marsh, M., Golledge, R., & Battersby, S. E. (2007). Geospatial concept understanding and recognition in G6–college students: A preliminary argument for minimal GIS. *Annals of the Association of American Geographers, 97*(4), 696–712.

Meyer, J. W., Butterick, J., Olkin, M., & Zack, G. (1999). GIS in the K-12 curriculum: A cautionary note. *The Professional Geographer, 51*(4), 571–578.

National Council for Geographic Education. (2014). *Webinar Archives*. Retrieved from http://www.ncge.org/webinar-archives-list

Pierson, M. E., & McNeil, S. (2000). Preservice technology integration through collaborative action communities. *Contemporary Issues in Technology and Teacher Education, 1*(1), 189–199.

Pomales-García, C., & Liu, Y. (2006). Web-based distance learning technology: The impacts of web module length and format. *The American journal of distance education, 20*(3), 163–179.

Pryor, C. R., & Bitter, G. G. (2008). Using multimedia to teach inservice teachers: Impacts on learning, application, and retention. *Computers in Human Behavior, 24*(6), 2668–2681.

Sui, D. Z. (1995). A pedagogic framework to link GIS to the intellectual core of geography. *Journal of Geography, 94*(6), 578–591.

Wilder, A., Brinkerhoff, J. D., & Higgins, T. M. (2003). Geographic information technologies + project-based science: A contextualized professional development approach. *Journal of Geography, 102*(6), 255–266.

Yu, J., Huynh, N. T., & McGehee, T. L. (2011). Vertical integration of geographic information sciences: A recruitment model for GIS education. *Journal of Geography, 110*(5), 191–199.

Chapter 11
Professional Development Focusing on Inquiry-Based Learning Using GIS

Lara M.P. Bryant and Tim Favier

Abstract Geographic inquiry projects with GIS make geography education more relevant and challenging, and provide opportunities to stimulate in-depth knowledge about geography, increase higher-order thinking and develop a wide range of technology skills. Providing professional development that ensures teachers know how to use GIS to support geographic inquiry is necessary. The professional development experiences should take into account the various competencies teachers need to design and conduct geographic inquiry projects using geospatial technologies. The TPACK framework focuses on teacher competencies and is valuable when designing professional development. Considering the TPACK framework, a successful strategy for teacher training is a collaborative inquiry model. When applied to GIS, the collaborative inquiry model is designed to overcome the common barriers to using GIS such as the lack of curriculum, support, and data. The model highlights the successful implementation of geographic inquiry using GIS within a school or district based collaborative team. This chapter explores didactic models for integrating GIS in inquiry projects; frameworks for the competencies that teachers need to design and conduct such projects; and successful strategies for training these competencies.

Keywords GIS • Professional development • Inquiry-based Learning

11.1 Introduction

As technology has become increasingly important in education, geography teachers and teacher trainers have become interested in the possibilities of using Geographic Information Systems (GIS) to enhance students' learning. GIS offers many

L.M.P. Bryant (✉)
Department of Geography, Keene State College, Keene, NH 03431, USA
e-mail: lbryant1@keene.edu

T. Favier
Faculty of Geosciences, Utrecht University, Utrecht, The Netherlands

Geography Department, Fontys University of Applied Sciences, Tilburg, The Netherlands
e-mail: t.t.favier@uv.nl

O. Muñiz Solari et al. (eds.), *Geospatial Technologies and Geography Education in a Changing World*, Advances in Geographical and Environmental Sciences,
DOI 10.1007/978-4-431-55519-3_11

opportunities to support geographic inquiry projects. Inquiry-based geography education with GIS makes learning more relevant and challenging, by connecting practice and theory, and stimulating the development of in-depth geographic knowledge and progression in a wide range of valuable skills such as geographic inquiry, spatial analysis, and critical thinking. However, integrating GIS into curriculum is not easy for teachers, because developing skills in teaching with GIS includes more than just learning to use the software. Teachers should learn to use GIS as a tool for developing students' geographic knowledge and skills by designing and conducting inquiry projects with GIS. In order to develop these skills, teachers need to be supported and provided with adequate professional development. This chapter explores didactic models for integrating GIS in inquiry projects; frameworks for the competencies that teachers need to design and conduct such projects; and a suggested model for training the necessary teacher competencies.

11.2 Supporting Inquiry-Based Geography Education with GIS

Due to media coverage regarding the status of geographic education, the general public often views geography as the memorization of topographical and factual knowledge about places and regions. This focus on discrete facts has been seen internationally in geographic education with inquiry and decision-making less emphasized (Gerber 2001). In the past few decades, more significance has been attached to the development of inquiry skills (Kent 2006). *Inquiry-based learning* (IBL) is a kind of learning which aims to stimulate progression in students' disciplinary subject knowledge and inquiry skills, thinking skills, and self-regulation skills by engaging in activities 'like researchers do' (van Joolingen et al. 2005). Recent reforms in geography education internationally have included movements toward IBL (Marsden 2003). However, the use of the IBL model of learning in geography education was described as early as 1921 by Smith, and mimics the scientific inquiry process. According to this model, inquiry is a cyclical process which consists of the following activities: (1) identifying the problem/hypothesis, (2) collecting data, (3) organizing data, (4) analyzing data or testing the hypothesis, and (5) determining and evaluating a solution (Asmussen and Buggey 1977; Fenton 1968; Moore and Wilcox 1932; Smith 1921). Figure 11.1 presents a model for geographic IBL with GIS, which illustrates the various activities, and the input and output of those activities (Favier 2011).

GIS is an ideal tool for supporting IBL in geography, as it allows teachers to design projects in which students explore spatial problems with digital maps, formulate questions about those problems, collect geodata in the field, visualize and analyze geodata in maps, and use these maps to answer their questions. The literature contains many enthusiastic descriptions of teachers using GIS to support IBL projects about natural disasters, pollution, solid waste, crime, health,

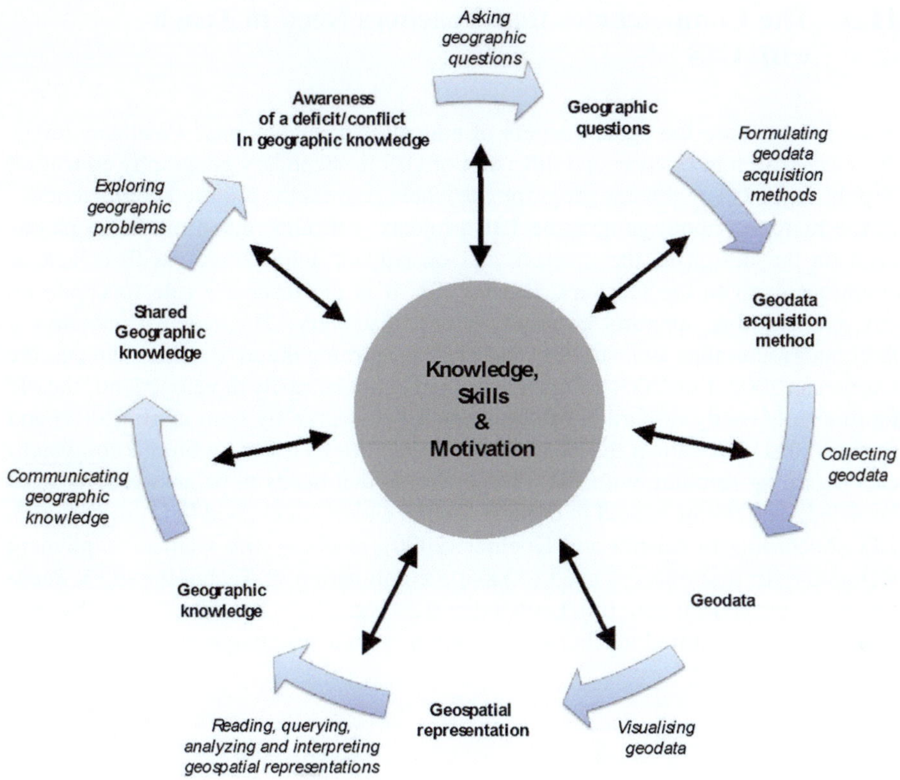

Fig. 11.1 A model for the geography IBL with GIS (Favier 2011)

recreation, market analysis, and political and cultural variability within regions (e.g. Audet and Ludwig 2000; Demirci et al. 2013; Engelhardt 2004; ESRI 2003; Falk and Nöthen 2005; Favier and Van der Schee 2012; Kerski 2003; Milson et al. 2012; Reitz 2005; Schleicher and Schrettenbrunner 2004; Sinton and Bednarz 2007; Unterthurner 2004).

Van Rens (2005) stresses the importance of paying attention to knowledge, skills, and motivation when designing and conducting IBL projects. Engaging in inquiry may stimulate progression in knowledge, skills, and motivation; but knowledge, skills, and motivation are also a precondition for engaging in inquiry. Therefore, GIS-based IBL projects can contribute to a progression in students' geographic knowledge and skills (geographic literacy) and students' motivation to learn about and solve problems in the world around us (the geographic drive). However, students should also have some background geographic knowledge and geographic inquiry skills, including GIS skills, and should be willing to engage in geographic inquiry.

11.3 The Competencies that Teachers Need to Teach with GIS

Since teachers are the "gate keepers of educational innovations" (Wallace 2004), the successful introduction and diffusion of GIS in secondary geography education largely depends on whether geography teachers possess the required competencies. In the literature about geographic IBL projects with GIS, much of the focus has been on the design of the projects and on student achievement, with a lack of attention given to the teachers. During IBL it is the teacher's role to create an environment that supports exploration and discovery. The teacher becomes a facilitator rather than an instructor, and student learning is active rather than passive (Anderson 2002; Luft 2001). This new role may not be easily developed, nor the old familiar role easily discarded by the teachers. Research by Lam et al. (2009) and Favier (2011) has shown that teachers often feel they lack the required competencies regarding teaching with IBL. The question that needs to be answered is what knowledge do teachers need in order to design and conduct good IBL projects with GIS. According to Mishra and Koehler (2006), teachers who want to implement technology in their classes need to have a combination of Technology (T), Pedagogy (P), and (A) Content (C), knowledge (K) (Fig. 11.2). The TPACK can be used to describe the required teacher knowledge base in a systematic way.

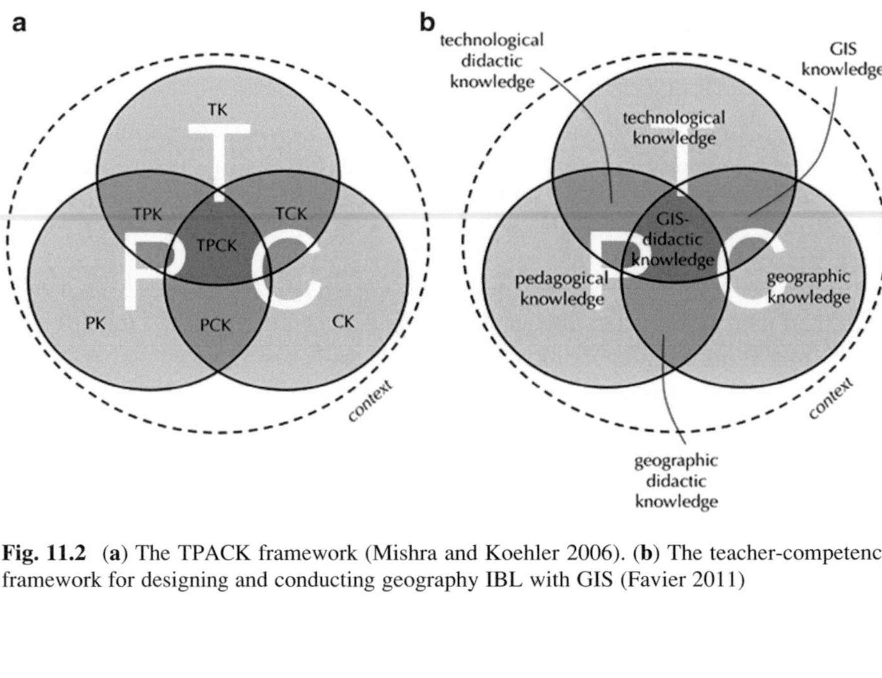

Fig. 11.2 (**a**) The TPACK framework (Mishra and Koehler 2006). (**b**) The teacher-competency framework for designing and conducting geography IBL with GIS (Favier 2011)

11.3.1 The Pedagogical and Content Components

It is clear that teachers need to have general pedagogical knowledge about the processes of teaching and learning for successful classroom management, lesson plan development and implementation, student evaluation, etc. This domain-generic knowledge is called Pedagogical Knowledge (PK) (Mishra and Koehler 2006, p. 1025). PK includes declarative knowledge about cognitive, social and developmental theories. It also includes procedural knowledge about applying these theories in classroom practice, and strategic knowledge about appropriate pedagogical interventions that are useful in specific situations.

Besides PK, teachers clearly also need disciplinary knowledge, or Content Knowledge (CK). Every discipline has core central ideas, concepts, and methods that teachers need to be fluent with in order to teach the subject: geography teachers should be geographically literate themselves. Shulman (1986) argued that combining PK and CK together does not make a teacher. Teachers also need to know how to teach the specific subject knowledge and inquiry methods knowledge of a domain, and therefore introduced the term Pedagogical Content Knowledge (PCK). This knowledge exists at the intersection of PK and CK.

For geography, PCK is *geographic-didactic knowledge*. Following Shulman (1986), Favier (2011) distinguished two sub-components of geographic-didactic knowledge. Knowledge in the first sub-component, '*geographic knowledge for use in educational settings*', includes knowledge about geographic issues and geographic inquiry methods that is transformed so that it becomes accessible for students. Teachers should know *how* to transform their own geographic knowledge to constructs for use in educational settings, and know *which* constructs for use in educational settings are suitable to reach specific learning goals. The second sub-component of geographic-didactic knowledge is '*student geographic inquiry learning processes in relation to tasks and coaching*' (Favier 2011). Knowledge in this sub-component includes: (1) declarative knowledge about how students learn in relation to task-design and coaching in geography classes, (2) procedural knowledge about how to design geographic inquiry tasks and coaching students when these tasks are conducted in classrooms, and (3) strategic knowledge about what kind of geographic inquiry tasks and coaching are suitable to reach specific learning goals.

11.3.2 The Technological Components

Teaching successfully with technology requires that teachers not only have knowledge in the pedagogical and content dimensions of education, but in technology as well (Mishra and Koehler 2006). Mishra and Koehler distinguish four kinds of technology-related knowledge that teachers should have. First, teachers need to have general Technological Knowledge (TK), which is knowledge about how to use standard software tools for text processing, spreadsheets, file management, and the Internet in non-educational settings. If they want to integrate such technologies in the classroom, they need to have Technological Pedagogical Knowledge (TPK): knowledge about how to use general technologies, such as digiboards and electronic learning environments, in the classroom. Geospatial technologies (GSTs), including GIS, are examples of domain-specific technologies, as they are especially suitable for analyzing *geographic* problems. In order to implement GIS in inquiry-based geography education, teachers should know how to use the software themselves to investigate geographic issues. This *GIS knowledge* can be seen as knowledge at the intersection of technology and content, and is therefore called Technological Content Knowledge (TCK). TCK includes declarative knowledge about the characteristics of geodata and the structure of GIS, procedural knowledge about how to apply GIS tools, and strategic knowledge about which sequence of tools should be applied in order to answer a specific geographic question.

However, even more important for successful integration of GIS in teaching is *GIS-didactic knowledge*, which is the overlapping knowledge of Technology, Pedagogy and Content (TPACK). Similar to geographic-didactic knowledge (PCK), TPACK can also be subdivided in two sub-components (Favier 2011). The first sub-component is '*GIS knowledge for use in educational settings*', which refers to suitable methods for data collection, data visualization and data analysis for student inquiry projects. The second sub-component is called '*student GIS-supported inquiry learning processes in relation to tasks and coaching*'. The declarative knowledge in this sub-component refers to knowledge about how students learn when they work on tasks with GIS. It includes declarative knowledge about frequently occurring problems, such as the fact that students rarely switch off irrelevant map layers, which makes it more difficult for them to analyse the representations. It also includes procedural knowledge about how to design good GIS tasks, and how to provide good instruction, support and reflection. In order to create viable and effective projects, teachers need to have sufficient declarative, procedural and strategic knowledge in every component of the TPACK framework (Favier 2011). MaKinster and Trautmann have developed a TPACK framework for GIS specific to the sciences, which include geographic content (2014).

11.4 Challenges Faced During Inquiry-Based Education When Teachers Lack the Required Competencies

Although many teachers believe GIS offers opportunities for inquiry-based geography education, many also agree that designing and conducting quality geographic inquiry projects is difficult (Lam et al. 2009; Favier 2011). Studies suggest the need for deepening teachers' understanding and application of general technologies (TK) (Bryant 2010), as well as their knowledge about how to teach with those technologies (TPK), before they can take steps in teaching with technologies (Kirschner and Davis 2003). With respect to teaching with GIS, research by Favier (2011) and Bryant (2010) suggests that it is not only teachers' limited GIS-knowledge (TCK) that forms an obstruction, but that teachers' geographic didactic knowledge (PCK) is also often insufficient for designing and conducting good geographic inquiry projects with GIS. For example, it was found that teachers had a difficult time connecting required curriculum to relevant local studies (Bryant 2010) and to structure geographic content and make it accessible for students (Favier 2011). It has also been found that teachers mistakenly perceived IBL as a process in which there are resources with answers to be found instead of a process in which answers can be derived though data collection and analysis (Bryant 2010; Crockett 2002; Gayford 2001). Therefore, teachers initially saw GIS as a useful resource, instead of as a tool for IBL and stimulating students' geographic thinking or inquiry skills. This implies that teachers need to have sufficient technological knowledge (TK), pedagogical knowledge (PK), and content knowledge (CK), before they can begin developing their technological-didactic knowledge (TPK), GIS-knowledge (TCK), and geographic-didactic knowledge (PCK), and that they should develop enough knowledge in these components before they can develop their GIS-didactic knowledge (TPCK) via designing, conducting, and evaluating simple inquiry projects with GIS.

11.5 A Recommended Model for Improving Teachers' Competencies

It is important to address teacher competencies in order to overcome the challenges teachers encounter when they implement IBL using GIS based projects. With many educational in-service programs focusing on classroom techniques, teachers have no example of how to implement the theoretical ideas behind geography IBL into practice (Lampert and Ball 1999; Thompson and Zeuli 1999). Teacher trainers should have the knowledge and abilities to model for teachers the appropriate processes in which teachers should engage their students (O'Hara and Pritchard 2008), including modeling inquiry-based lessons (Luft 2001). Therefore, the trainers should be facilitators themselves, not instructors. The best way to model the inquiry process to teachers is to immerse them in the process of inquiry themselves (Supovitz and Turner 2000) and let them structure the geographic issues themselves, and explore how to translate the content to constructs for use in educational settings. Trainings should focus on those elements of GIS that can ensure achievement in educational objectives such as in-depth geographic knowledge, inquiry skills, critical thinking skills, and geographic thinking skills, and should aim to stimulate progression in knowledge in all components of the TPACK framework.

11.5.1 *The Collaborative Inquiry Model*

The Collaborative Inquiry Model (CIM) (Bryant 2010) provides guidelines for training teachers' competencies in designing and conducting IBL projects with GIS. The model is based on the following principles: (1) follow the philosophy of social constructivism and allow the participants to work collaboratively, (2) create a network of teachers, peers and professionals in the field for ongoing support, (3) provide immediate connection with existing curriculum and content, (4) use local relevant examples that provide a framework upon which to construct global knowledge, and (5) stimulate critical reflection. When these five principles are included in the conceptual framework of professional development, it increases the effective implementation of GIS in the classroom.

In a phenomenological research project, teachers designed, tested and evaluated IBL projects with GIS together by following the principles of CIM (Bryant 2010). The research focused on teachers who participated in a four day summer institute, Geographic Inquiry using GIS which also included two follow-up sessions during the school year. In order to foster the development of a collaborative professional community, the institute targeted teachers from a single urban school district in Texas, United States. Eleven middle and high school teachers and a district instructional technologist participated in the study. During the institute, teachers explored the use of online interactive mapping systems, ArcExplorer, and ArcGIS 9.2. Institute activities included the use of *Mapping our World Using GIS* (Palmer et al. 2008), data collection in the field, and partnerships with local agencies to inform the development of the inquiry-based lessons.

Before beginning, teachers were given time to determine standards and objectives they already teach that would be suited to use with GIS, and plan accordingly (McClurg and Buss 2007). Then they proceeded to work through each step of the inquiry process, and explored how the various components could apply to their specific curriculum. They were reflective throughout the process and were critical about their inclusion of GIS into the curriculum, learning the difference between using the program solely for technology's sake and using it as a tool for teaching geographic subject knowledge and inquiry skills.

11.5.2 *Impacts of the Collaborative Inquiry Model on Teachers*

By guiding teachers through the geographic inquiry process using the CIM, there was increased meaningful use of GIS during their development of inquiry-based projects (Bryant 2010.) First, teachers reflected upon who should ask the questions during the inquiry process. Some teachers arrived at the same conclusions as Smith (2005) that independent exploration and questioning of geographic data makes using the program more motivating to the student, as it encourages students to be "inquisitive" and "open-minded". So paying attention to the 'geographic drive' is fruitful. Others chose to guide or model the questioning process to the students. Both of these methods are very different in regard to developing the skills necessary for students to learn how to ask their own geographic questions and are dependent on the development stage of the learners themselves. Together teachers discovered the variations in teaching that would lead to the desired objective of students eventually being able to formulate questions themselves.

Second, teachers examined the importance of data acquisition in developing IBL projects using GSTs. Acquiring data can be as simple as turning on a layer in an online database, or more advanced by downloading the layer for use in the desktop software, joining tables to shapefiles, or collecting the data in the field. The inquiry process depends not only on the ability to collect and gather geospatial data, but the accessibility of that data as well developing strategies of data collection. In order to be useful, the data need to be connected to existing curricula, and broad enough to allow students to pursue lines of inquiry (Liu and Zhu 2008). Providing a regional data set for teachers during trainings has been a successful strategy (McClurg and Buss 2007). Community-based projects involving student-collected data in citizen science projects have also been very successful. Depending on the objectives of the lesson, teachers will weigh the value of data collection differently (Bryant 2010). Teachers who focused on the development of research skills wanted students to collect their own data, while teachers who wanted primarily to strengthen questioning strategies and analysis placed less emphasis on data collection.

Teachers also improved their GIS skills regarding organizing and analyzing data. This improvement was evident in their changed expectations from having students perform general activities creating maps to requiring that students conduct analytical tasks such as ranking countries, classifying data to make thematic maps or embedding evidence such as photos, tables or charts. Most of the teachers still expected the students to conduct visual pattern analysis. This fact suggests that either the skill level of the teachers did not improve to the level needed to include using GIS as part of the analysis process, or that the visual analysis and development of student spatial skills was the intent of the developed curriculum. Finally, the most noteworthy change in teacher expectations after the CIM experience was the fact that the teachers expected the students to use the program to find data that would support the conclusions, predictions, or answers the students derived instead of simply expecting the students find an answer (Bryant 2010). This is fundamental to IBL for students and the reason why collaborative inquiry models for teacher education are promising.

11.6 Recommendations for Teacher Professional Development

Teacher education should not only focus on training the technical knowledge (general technological knowledge, GIS knowledge and GIS didactic knowledge), but also on the geographic didactic knowledge. Teachers need knowledge in all components of the GIS TPACK framework. Working collaboratively, to develop knowledge within the TPACK framework is important for teacher success. Therefore, schools should take advantage of existing teaming and collaboration already in place in many school structures to improve successful implementation of GIS in the classroom (Bryant 2010). Finally, teachers benefit from the time to process their own student needs, examine how GIS can support mandated priorities and reflect upon their teaching practices in order to design curriculum accordingly.

References

Anderson, R. D. (2002). Reforming science teaching: What research says about inquiry. *Journal of Science Teacher Education, 13*, 1–12.

Asmussen, D., & Buggey, J. (1977). Teaching geography through inquiry. In G. A. Manson & M. K. Ridd (Eds.), *New perspectives on geographic education: Putting theory into practice* (pp. 99–124). Dubuque: Kendall/Hunt Publishing Company.

Audet, R. H., & Ludwig, G. S. (2000). *GIS in schools*. Redlands: ESRI Press.

Bryant, L. (2010). *A phenomenological study of a collaborative inquiry model for teaching educators to use geographic information systems*. Ph.D. dissertation, Texas State University, San Marcos, Texas.

Crockett, M. D. (2002). Inquiry as professional development: Creating dilemmas through teachers' work. *Teaching and Teacher Education, 18*, 609–624.

Demirci, A., Karaburun, A., & Ünlü, M. (2013). Implementation and effectiveness of GIS-based projects in secondary schools. *Journal of Geography, 112*(5), 214–228.

Engelhardt, G. (2004). Kiesabbau am Mittleren Oberrhein: Unterrichtsprojekt im Geographieunterricht der Kursstufe. *Praxis Geographie, 2*, 24–26.

ESRI. (2003). *Making community connections*. Redlands: ESRI Press.

Falk, G. C., & Nöthen, E. (2005). *GIS in der schule, potenziale und grenzen*. Berlin: Mensch and Buch Verlag.

Favier, T. T. (2011). *Geographic Information Systems in inquiry-based secondary geography education*. Ph.D. Dissertation, Vrije Universiteit, Amsterdam, The Netherlands.

Favier, T. T., & Van der Schee, J. A. (2012). Exploring the characteristics of an optimal design for inquiry-based geography education with Geographic Information Systems. *Computers and Education, 58*, 666–667.

Fenton, E. (1968). The problem or critical analysis approach. In J. W. Morris (Ed.), *Methods of geographic instruction* (pp. 165–176). Waltham: Blaisdell Publishing Company.

Gayford, C. (2001). Education for sustainability: An approach to the professional development of teachers. *European Journal of Teacher Education, 24*, 313–327.

Gerber, R. (2001). The state of geographical education in countries around the world. *International Research in Geographical and Environmental Education, 10*, 349–362.

Kent, A. (2006). Innovation, change and leadership in geography departments. *Geography, 91*, 117–126.

Kerski, J. J. (2003). The implementation and effectiveness of Geographic Information Systems technology and methods in secondary education. *Journal of Geography, 102*, 128–137.

Kirschner, P. A., & Davis, N. (2003). Pedagogic benchmarks for ICT teacher education. *Technology, Pedagogy and Education, 12*, 149.

Lam, C. C., Lai, E., & Wong, J. (2009). Implementation of geographic information system (GIS) in secondary geography curriculum in Hong Kong: Current situations and future directions. *International Research in Geographical and Environmental Education, 18*, 57–74.

Lampert, M., & Ball, D. (1999). Aligning teacher education with contemporary K-12 reform visions. In L. Darling-Hammond & G. Sykes (Eds.), *Teaching as the learning profession: Handbook of policy and practice* (pp. 33–53). San Francisco: Jossey-Bass Publishers.

Liu, S., & Zhu, X. (2008). Designing a structured and interactive learning environment based on GIS for secondary geography education. *Journal of Geography, 107*, 12–19.

Luft, J. A. (2001). Changing inquiry practices and beliefs: The impact of an inquiry-based professional development programme on beginning and experienced secondary science teachers. *International Journal of Science Education, 23*, 517–534.

MaKinster, J., & Trautmann, N. (2014). The nature of teacher knowledge necessary for the effective use of geospatial technologies to teach science. In J. MaKinster, N. Trautmann, & M. Barnett (Eds.), *Teaching science and investigating environmental issues with geospatial technology: Designing effective professional development for teachers* (pp. 333–353). Dordrecht: Springer.

Marsden, B. (2003). Geography curriculum planning in evolution: Some historical and international perspectives. In R. Gerber (Ed.), *International handbook on geographical education* (pp. 141–157). Dordrecht: Kluwer Academic Publishers.

McClurg, P. A., & Buss, A. (2007). Professional development: Teachers' use of GIS to enhance student learning. *Journal of Geography, 106*, 79–87.

Milson, A. J., Demirci, A., & Kerski, J. J. (Eds.). (2012). *International perspectives on teaching and learning with GIS in secondary schools*. New York: Springer.

Mishra, P., & Koehler, M. J. (2006). Technological pedagogical content knowledge: A new framework for teacher knowledge. *Teachers College Record, 108*, 1017–1054.

Moore, C. B., & Wilcox, L. A. (1932). *The teaching of geography*. New York: American Book Company.

O'Hara, S., & Pritchard, R. H. (2008). Meeting the challenge of diversity: Professional development for teacher educators. *Teacher Education Quarterly, 35*, 43–61.

Palmer, A. M., Palmer, R., Malone, L., & Voigt, C. L. (2008). *Mapping our world using GIS*. Redlands: ESRI Press.

Reitz, V. (2005). Nutzungsconflicte im Nationalpark Schleswig-Holsteinisches Wattenmeer, Inhaltliches Lernen mit SchulGIS. *Geographie Heute, 233*, 41–44.

Rens, L. V. (2005). *Effectief scheikundeonderwijs voor 'leren onderzoeken' in de tweede fase van het vwo. Een chemie van willen, weten en kunnen*. Amsterdam: Vrije Universiteit.

Schleicher, Y., & Schrettenbrunner, H. (2004). Schädigungen von Bäumen auf dem Schulgelände, Eind Unterrichtbeispiel erstellt mit SchulGIS. *Praxis Geographie, 2*, 21–23.

Shulman, L. S. (1986). Those who understand: Knowledge growth in teaching. *Educational Researcher, 15*, 4–14.

Sinton, D. S., & Bednarz, S. W. (2007). About that G in GIS. In D. S. Sinton & J. J. Lund (Eds.), *Understanding place: GIS and mapping across the curriculum* (pp. 19–34). Redlands: ESRI Press.

Smith, E. E. (1921). *Teaching geography by problems*. Garden City: Doubleday, Page & Company.

Smith, J. S. (2005). Flow theory and GIS: Is there a connection for learning? *International Research in Geographical & Environmental Education, 14*, 223–230.

Supovitz, J. A., & Turner, H. M. (2000). The effects of professional development on science teaching practices and classroom culture. *Journal of Research in Science Teaching, 37*, 963–980.

Thompson, C. L., & Zeuli, J. S. (1999). The frame and the tapestry: Standards-based reform and professional development. In L. Darling-Hammond & G. Sykes (Eds.), *Teaching as the learning profession: Handbook of policy and practice* (pp. 341–375). San Francisco: Jossey-Bass Publishers.

Unterthurner, H. (2004). GIS-Projekt Hotelführer Meran. *Praxis Geographie, 34*(2), 28–31.
van Joolingen, W. R., deJong, T., Lazonder, A. W., Savelsbergh, E. R., & Manlove, S. (2005). Co-Lab: Research and development of an online learning environment for collaborative scientific discovery learning. *Computers in Human Behavior, 21*, 671–688.
Wallace, R. M. (2004). A framework for understanding teaching with the Internet. *American Educational Research Journal, 41*, 447–488.

Part IV
Evaluation and Assessment on Geospatial Practices

Chapter 12
The Effectiveness of Geospatial Practices in Education

Ali Demirci

Abstract Geospatial technologies (GST) has long been used in education in many countries. Teachers and students from around the world are using desktop and Web GIS, Global Positioning Systems (GPS), Google Earth, and many other location based services available from the Internet in teaching and learning various school subjects such as geography, environmental sciences, social sciences, history, biology, and mathematics. Potentials of GST for teaching and learning have been identified in many studies. GST provides teachers with a dynamic platform where they can incorporate inquiry-based, student-centered and many other constructivist teaching methods in their lessons. Geospatial practices has a great potential for students to equip them with versatile knowledge and skills as well as to improve their achievements in lessons. Although a great majority of the literature expresses many different benefits of GST for education, some studies raise concern about its effectiveness by addressing the need to find the proper methods for its implementation. This chapter first evaluates the potentials of GST for teaching and learning and then discusses whether geospatial practices are actually effective mainly for secondary education by looking at different concerns and questions raised in the related literature.

Keywords Geospatial technologies • Effectiveness • Geography education

12.1 Introduction

Geospatial technologies (GST) have long been used in education in many countries. Teachers and students from around the world are using desktop and web GIS, Global Positioning Systems (GPS), Google Earth and many location based services available from the Internet in the teaching and learning of school subjects such as geography, environmental sciences, social sciences, history, biology and mathematics. The potential of GST for teaching and learning have been identified in many

A. Demirci (✉)
Department of Geography, Fatih University, Istanbul, Turkey
e-mail: ademirci@fatih.edu.tr

O. Muñiz Solari et al. (eds.), *Geospatial Technologies and Geography Education in a Changing World*, Advances in Geographical and Environmental Sciences,
DOI 10.1007/978-4-431-55519-3_12

studies. Although a great majority of the literature expresses many different benefits of GST for education, some studies raise concerns about its effectiveness by addressing the need to find the proper methods for its implementation.

12.2 Potentials of GST for Teaching and Learning

The benefits of GST for teaching and learning have been identified in many studies since the beginning of the 1990s. Many studies addressed the supporting role of GST as a versatile tool for teaching and learning geography (Bednarz and Van der Schee 2006; Demirci 2011; Kim et al. 2011; Patterson et al. 2003; Wang and Chen 2013). Although it can even be used in traditional educational settings (Lidstone and Stoltman 2006), GST changed the landscape of teaching and learning (Alibrandi 2003), and transformed it into an environment in which constructivist educational strategies are applied (Bednarz and Ludwig 1997; Doering and Veletsianos 2008; Kerski et al. 2013). Instead of being passive receivers of information, constructivist approaches and methods with GST make students active explorers of their own understanding (Huang 2011) and allow them to learn through their own experiences (Bednarz 2004).

GST provides students with a large amount of information and many mapping tools (Huang 2011) for geographic data analysis, exploration and visualization with which students gather and analyze information (Liu and Zhu 2008) to ask and answer geographic questions (Shin 2006). Geospatial practices turn students into researchers (Baker and White 2003) by helping them to visualize and examine geographic patterns in their data (Breetzke et al. 2011), and relationships between and among spatial phenomena (Stoltman and De Chano 2003). Practices with GST increase student-map interactivity, which enables students to see the possibility of discovering unknowns (Wiegand 2003). A diverse set of activities such as gathering, storing, visualizing, querying, analyzing and managing data with various GST supports many constructivist educational strategies, approaches and methods. According to Liu and Zhu (2008, p. 14), "these tools [GIS technology] can support geographic inquiry by allowing learners to formulate geographic questions or hypotheses, access and obtain geographic data from multiple sources, present geographic data and information in forms of maps, images, tables, and charts, explore the data through carefully constructed queries, and analyze the data to answer the questions or draw conclusions." As addressed in many other studies, GST supports issue-based, student-centered and standard-based education (Kerski 2003), encourages problem, inquiry and project based learning (Akerson and Dickinson 2003; Favier and Van der Schee 2012; Lidstone and Stoltman 2006; Meyer et al. 1999; Nielsen et al. 2011), facilitates collaborative work, individual learning (Baker and White 2003; Keiper 1999) and assists to create inductive learning environments (Milson and Earle 2008).

Practicing with GST to enhance teaching and learning in geography can have broad educational benefits for students, which can be classified in three different categories in this chapter. GST helps students to (1) develop skills, (2) provide knowledge and (3) gain motivation, attitude and understanding. One of the most important reasons for practicing with GST in education is that it provides students with many valuable skills. Apart from various personal, entrepreneurial and marketable skills that enhance students' future careers (Goldstein and Alibrandi 2013) and contribute to improving youth employability (Shin 2006), teaching with GST encourages critical, effective and scientific thinking in education (Akerson and Dickinson 2003; Bevainis 2008; Goldstein and Alibrandi 2013; Roulston 2013). As addressed in a growing number of studies, GST also has great potential to develop high-order-thinking skills (Doering and Veletsianos 2008; Kerski 2003; Linn et al. 2005; West 2003; Wilder et al. 2003), enabling students to apply, analyze, evaluate and create information rather than merely memorizing it (Liu et al. 2010). Spatial thinking is also among the most cited skills that GST develops, especially for geography education (Audet and Abegg 1996; Bednarz 2004; Biilmann 2001; DeMers and Vincent 2008; Lee and Bednarz 2009; Kerski 2008; Wiegand 2001). Being an important skill for everyday life, spatial thinking is used to solve problems by analyzing the spatial relationships of objects and places with reference to locations, distances, directions, shapes and patterns (Kidman and Palmer 2006). Practices with GST allow students to perform functions such as spatial querying, statistical analysis and visualization, which facilitate students' manipulating, querying, analyzing, summarizing and editing spatial data (Goldstein and Alibrandi 2013). All these functions of GST help students to think spatially (Lee and Bednarz 2009), to ask spatial questions (Nellis 1994), to gain spatial awareness (West 2008) and, finally, to solve spatial problems (Audet and Paris 1997; Demirci et al. 2013a). By considering all the skills, whether mentioned here or not, we can easily say that many different in and out-of-class practices involving GST helps students to think geographically (Baker et al. 2009; Shin 2006), and to ask and to answer geographic questions by acquiring, organizing and analyzing geographic information, which are the key issues in secondary school geography education (Keiper 1999; Schultz et al. 2008).

The practice of GST has a great potential to enhance geographic learning and to improve geographic literacy (Benimmas et al. 2011; Favier and Van der Schee 2012; Liu and Zhu 2008; Shin 2007; Wechsler and Pitts 2004). Students practicing GST can understand geography more efficiently (Demirci 2008) by exploring geographic issues and problems (Bednarz and van der Schee 2006; Lemberg and Stoltman 2001; Liu and Zhu 2008) with real-world relevance to the subject (Baker et al. 2009). Shin (2006) studied the use of GIS with primary school students and concluded that it improved geographic literacy better than other methods. In their quantitative study, Wechsler and Pitts (2004) practiced GST with high school students and asserted that the application significantly increased students' geographic knowledge. In another study, Milson and Earle (2008) observed an enhancement in geographic learning when they used an Internet-based GIS with their students. GST can also improve students' abilities to carry out location-based scientific research (Baker and White 2003) and provide them with important tools to explore and study local issues and environments (Bednarz 2004; Lemberg and

Stoltman 2001). Geospatial practices also allow students to study and understand local, regional and global geographical issues and problems and thereby enhances students' achievement in geography lessons (Demirci 2008; Goldstein and Alibrandi 2013; Wechsler and Pitts 2004).

Skills and knowledge are not the only attributes that can be developed and enhanced by the practice of GST. As stated in many studies, GST has an even greater potential to motivate students to learn and make them more interested in lessons. Goodchild and Kemp (1990) addressed this more than two decades ago by stating that GIS helps enhance students' interest in geography and motivates them toward careers in science and engineering. Many other studies consolidated this and presented that learning with GST improves students' attitudes, motivation, self-efficacy and enthusiasm in geography lessons (Baker and White 2003; Demirci 2008; Demirci et al. 2013b; Doering and Veletsianos 2008; Kerski 2003; Nielsen et al. 2011). If applied with proper methods, GST may also make students more responsible and sensitive toward local/global issues and problems (Demirci et al. 2013b; Keiper 1999). The author of this chapter worked together with 124 students from three public schools in Turkey for nearly a year in nine different GIS-based projects that were related to local communities, and found that the students' sensitivity towards society and its problems increased (Fig. 12.1) (Demirci et al. 2013b). In a similar study, Milson and Earle (2008) stated that Internet-GIS projects can benefit students with cultural awareness and empathy for distant others.

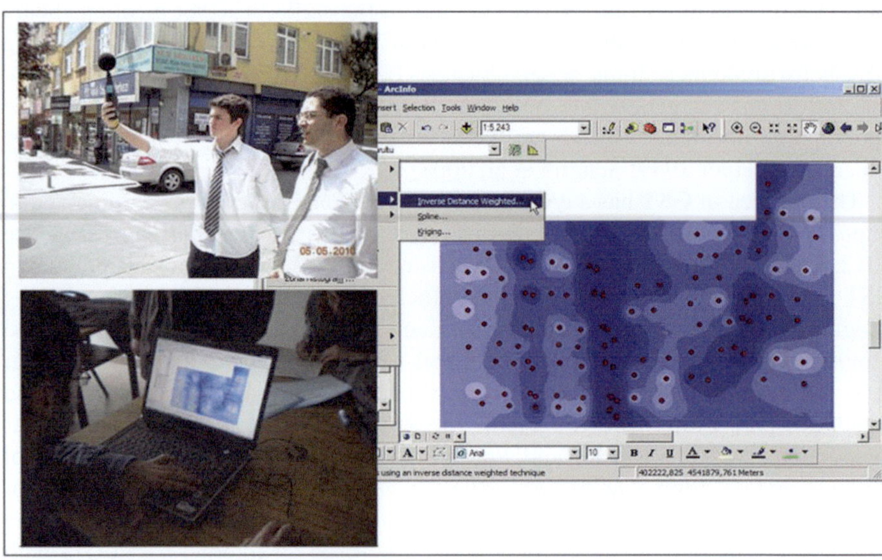

Fig. 12.1 High school students in Turkey are measuring noise pollution with GIS in their school district

12.3 Are Geospatial Practices Actually Effective in Teaching and Learning?

Due to its versatile benefits for teaching and learning, the practice of GST in education has expanded from a few countries, such as the USA, the UK and Canada, to many other countries around the world, first with GIS started in the early 1990s, and then began to include other digital technologies such as GPS, web-based mapping/GIS systems and many other location based applications available from the Internet (Kerski et al. 2013). Teachers and students today are utilizing GST in and out of classroom settings, in many countries across the six continents, including Norway, Germany, Austria, Uganda, Rwanda, Colombia, Chile, Taiwan, Singapore and Japan (Bevainis 2008; Lay et al. 2013; Milson et al. 2012a). Although the practices of GST has been gradually spreading around the world, its use in the classrooms lags far behind what researchers had hoped for more than a decade ago (Doering and Veletsianos 2008; Kim et al. 2011; Roulston 2013). Kerski et al. (2013) confirmed this when they studied the use of GIS in 33 countries, in a recent study. They found that the current global landscape of GIS remains small for secondary education. GIS has still not become a widely used, effective teaching tool, even in countries such as the US and the UK, which pioneered the use of GIS in education more than two decades ago (Bednarz and van der Schee 2006; Henry and Semple 2012; Kerski 2003).

The slow rate of practicing GST at the secondary school level has been attributed to many different challenges and obstacles in different studies (Baker 2005; Baker et al. 2009; Bednarz and Audet 1999; Chun and Hong 2007; Kerski 2003; Lidstone and Stoltman 2006; Milson and Earle 2008). Milson and Kerski (2012) identified these challenges as technological, pedagogical, administrative and curricular hurdles (Fig. 12.2). Technological challenges are related to the availability of data, software, computer and Internet infrastructure. Teachers' lack of skills, knowledge and experiences about GIS, and lack of motivation to use it in their lessons constituted pedagogical hurdles. Administrative obstacles were mainly related to whether or not the conditions in schools and in the education systems were favorable toward the use of GST in lessons (e.g., school managers' attention towards using GIS). Curricular impediments were the lack of strong subjects or curriculums to host GIS in schools (Demirci et al. 2013b). Some studies attributed the slow take-up rate of GST to other reasons, such as lack of research showing the effectiveness of this technology in education. According to Kerski (2003), the effectiveness of GIS in teaching and learning is unclear: this is among the reasons behind the low interest in GIS. Baker and Bednarz (2003) have also identified a lack of sufficient research on the effectiveness of GIS as an important obstacle preventing the use of GIS in schools. After all, the studies have been carried out over, roughly, the last three decades; can we not say very strongly that GST is effective teaching and learning tool for secondary schools? We need to look at the literature in order to elaborate on the possible answers.

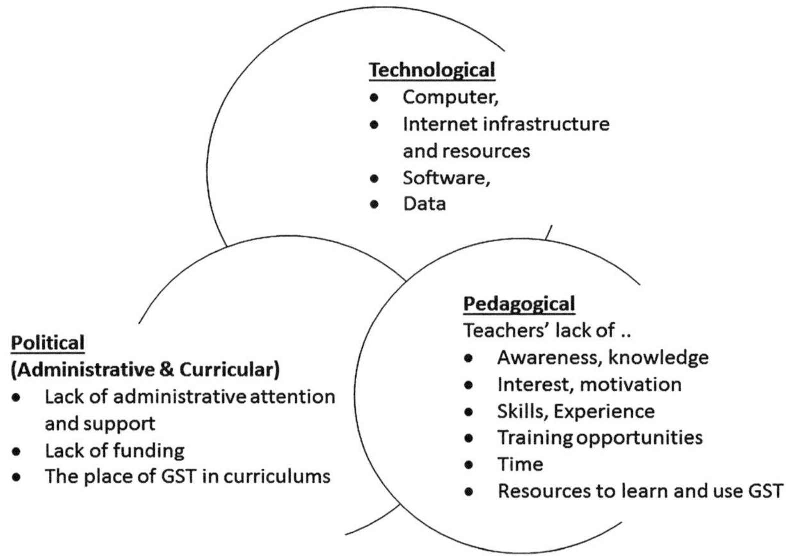

Fig. 12.2 Conditions affecting the use of GST in education

The educational benefits of practicing GST in secondary education began to be discussed in research papers at the beginning of the 1990s (Patterson et al. 2003); however, empirical data showing the effects of GIS on geographic learning, motivation, spatial ability and problem solving started to emerge by the late 1990s (Huynh 2009). There is not a study in the literature advocating that GST is useless for education. In general, there are two groups of studies concerning the effectiveness of GST for teaching and learning. The first group states that GST is effective in teaching and learning, despite reservations about a proper implementation method being raised. Although the majority of these studies are qualitative and based on theory, some important quantitative studies have been carried out, especially in recent years. The majority of these studies indicated that GST was effective, especially in making geography lessons more visual, student-centered and desirable by increasing students' achievements (Demirci 2008, 2011; Goldstein and Alibrandi 2013; Keiper 1999; Kerski 2003; Meyer et al. 1999; Patterson et al. 2003; Shin 2007; West 2003).

Audet and Abegg (1996) conducted one of the early studies to understand the effects of teaching with GST. After using GIS in a pilot study with high school students, they found that GIS was helpful for students when developing problem-solving abilities. Keiper (1999) also conducted a study in the late 1990s to understand the cognitive implications of GIS use in education. After conducting a GIS learning project with his students, Keiper stated that the project dramatically shifted the study of geography from memorization of places to the practice of geography skills; therefore, it encouraged the use of geographic knowledge.

The studies aiming to measure the effectiveness of geospatial practices for teaching and learning increased in number in the early 2000s. Baker and White (2003) developed and implemented a project based learning unit with two groups of students: one with collaborative GIS, the other with paper maps from an eighth grade Earth science lesson. They found a significant improvement in attitudes toward technology and in geographic data analysis for students who used GIS. Kerski (2003) carried out another important quantitative study almost at the same time. Kerski developed 12 geography lessons and implemented them with two methods: one with GIS, the other with traditional print materials. He found that the practices with GIS had a significant effect on student performance, increased students' test scores and improved students' abilities to synthesize, identify and describe reasons for human and physical patterns. In the same study, Kerski also concluded that GIS practices fostered students' higher-order analytical and synthetic thinking.

The effectiveness of practicing GST has been analyzed in more detail in many other recent studies. Liu et al. (2010) evaluated problem-based learning using GIS technology in a Singapore secondary school with students in experimental and control groups. They observed that students in the control group showed memorization skills, while students in the experimental group demonstrated higher-level cognitive learning skills, especially analytical and evaluation skills. In another recent study, Perkins et al. (2010) stated that a three-day GIS/GPS curriculum significantly increased students' spatial awareness. Goldstein and Alibrandi (2013) carried out quantitative analyses on standardized test scores of two groups of middle school students, with and without GIS instructions, and found out that GIS instructions significantly affected students' achievement on reading scores and on final course grades in science and social studies.

The second view in the literature concerning the effectiveness of GST usually raises concern that the studies and experiences of teaching with GST have not targeted higher order thinking skills especially in secondary education. Therefore, the effectiveness of GST has not yet been proven. There is a need to develop further research to determine whether GIS and other GSTs are actually effective for teaching and learning. We need to look at the literature from a historical perspective in order to understand the concerns raised in relation to their underlying reasons. Since there was not enough evidence, if any, showing that GIS was an efficient tool to enhance education, the studies published in the 1990s addressed the lack of empirical data to aid in understanding the real effect of GIS in education. Audet (1993) argued that GIS should be used in teaching and learning environments only if it could be proved that it enhanced the way students visualized and interpreted information. Bednarz and Ludwig (1997) raised the same issue by saying that clear evidence was needed to understand if GIS was an effective teaching and learning tool in order to persuade teachers of its value. The studies carried out until the early 2000s were mainly based on intuition or assumptions that the use of GST supported constructivist learning environments; however, they did not provide enough pedagogical evidence concerning the effectiveness of GST as an educational tool (Biilmann 2001; Keiper 1999; Lemberg and Stoltman 2001).

One of the most critical studies came from Bednarz (2004), where it was questioned whether GIS was a tool that would support geography and environmental education. This study was published at nearly the same time as other empirical studies that evaluated the effectiveness of GIS for teaching and learning, such as the ones conducted by Kerski (2003) and Baker and White (2003). Bednarz, in the study, addressed questions regarding the benefits, rationales and necessities of teaching with GIS in education. For example, what insights does GIS allow that the other ways of learning do not? This was asked at the first conference on the educational application of GIS, organized in the US in 1994, yet, the same questions remained unanswered a decade later in 2003. By drawing attention to pedagogical issues, Bednarz (2004, p. 198), in the same study, said "we cannot afford to continue to assume that, simply by doing GIS, students will recognize or learn cognitive mapping processes, spatial analysis or spatial thinking".

Different questions and concerns were raised in the following years regarding the effectiveness of GST for education. In their editorial note in the Journal of Geographical Education, International Research in Geographical and Environmental Education, Lidstone and Stoltman (2006, p. 206) asked, "how much GIS should students know, how should they use it, and how long will the operational skills persist in the minds of the learners?" In his Ph.D. dissertation, West (2008, p. 96) repeated almost the same concerns raised in the previous studies by saying, "whether or how using GIS enables students to attain the goals of geography remains largely unknown". In another doctoral study completed in 2009, Huynh (2009) addressed the fact that the research carried out up to that point had usually focused on geographic knowledge, skills, problem solving and attitudes; however, they missed some important areas such as the fundamental knowledge and skills needed for effective GIS use, which is a point still needs further study today to clarify.

Nearly all the studies that raised concerns and questions about the effectiveness of geospatial practices in education actually supported the general view that GIS and other geospatial technologies have great potential and many possible benefits for teaching and learning if they are used with proper methods. Bednarz and van der Schee (2006, p. 203) expressed this as "skeptical enthusiasm", by saying that they had been enthusiastic about the potential of GIS, but unsure about its fit with the traditional geography curriculum. The concerns and questions raised in these studies generally stem from a search to understand the true and specific values and benefits of GST compared to other technologies and methodologies for teaching and learning. As Doering and Veletsianos (2008) addressed, if the practices of GST enable learners only to employ and present data passively, there will not be much difference in the learning process when similar actions were applied with a different method or technology. A paper-based GIS exercise conducted in South Africa was found to be an adequate alternative to contemporary computerized GIS teaching methods (Breetzke et al. 2011). However, as emphasized by Shin (2006), GST provides students a platform upon which they can interact with data in a dynamic environment where they can manipulate and experiment in a way that would be difficult to do with other types of materials.

12.4 Conclusion

Another decade has passed since Bednarz (2004) raised questions about the effectiveness of teaching with GIS, and we still cannot say that we have answered all of those questions. New questions and concerns appear while GST and our needs in education to use it change and diversify. However, what we can say today is that many more researchers from around the world have been studying the use of GST in the teaching and learning of different subjects at schools. An important difference is that recent and new studies are mainly focusing on how we can benefit more efficiently from continuously evolving and changing GST, rather than discussing the effectiveness of these educational tools.

The answer of the question asked in the title of the previous section as "Are geospatial practices actually effective in teaching and learning?" is simply "yes". However, we need to define what we mean by effectiveness in order to have more detailed and satisfactory answers. Many versatile benefits of GST for teaching and learning described throughout this paper indicate that GST has an effective tool in many different ways; therefore, many different measurements should be taken into consideration to determine its true and detailed effectiveness. A study may not confirm a meaningful contribution of GIS to students' test scores, because using any inquiry based method like GIS does not fit very well with standardized tests. Even in this case, we cannot easily say that GIS was not effective, because it might have been effective in many other ways, such as in enhancing students' spatial thinking and critical thinking skills, motivating them to raise questions and search for answers by working with data, connecting them with community and global issues, helping them to choose a career, and even helping them to stay in school and graduate. Therefore, before studying the effectiveness of GST, we first need to determine the specific area in which we are seeking its contribution, and then we need to have an appropriate assessment methodology to measure its effectiveness. Some of the recent studies developed and tested different assessment methodologies to measure the effectiveness of GIS in specific areas such as in enhancing spatial thinking skills (Lee and Bednarz 2012).

The slow adoption of utilizing GST in classrooms such as GIS, GPS, Google Earth, and other location based services provided from the Internet does not mean that these technologies are not effective for teaching and learning, as countless examples from around the world, some of which have been outlined in this chapter, demonstrate successful stories. However, among many different challenges, the most important problem seems to stem from the traditional teaching strategies that we are accustomed to using in our classrooms and never want to abandon. We would like to use GST effectively for analysis, synthesis and evaluations that constitute higher order thinking skills; however, we do not question enough how much our current geography, or other lessons, support these important thinking skills with their content, curriculum, teaching methods and resources without the involvement of GST. As Bednarz and van der Schee (2006) said, we have to work together with teachers and students to conduct our lessons in different ways through inquiry and problem based learning in order to foster such skills. However, the majority of our teachers with poor preparation in geography lack the fundamental concepts related to asking and answering important spatialized questions (Bednarz and van der Schee 2006). For this important reason, many studies emphasize teacher-centered problems as the most important obstacles when using GST in classrooms effectively (Bednarz 2004; Bednarz and van der Schee 2006; Bevainis 2008; Kim et al. 2011). The author of the chapter experienced and observed this problem personally in one of his studies: the teachers who remained only observers and did not want to participate in a GIS based project conducted in their schools were the ones who mainly conducted their lessons via traditional methods, including the use of textbooks and descriptions (Demirci et al. 2013b).

Rapid developments in science and technology have been transforming geospatial technologies quickly. Every passing day, more and more people are becoming involved in many different forms of these technologies in their routine lives: commonly through their tablets, computers and smart-phones with an Internet connection. This development has already created a contrast between teachers and students in the classrooms in terms of their understanding and use of new technologies, which is described as the "digital divide" (Milson et al. 2012a, b) or a clash between the "digital" students and the "analogue" teachers (Svatonova and Mrazkova 2010). Geospatial technologies are most likely to become one of the most widely used educational tools in secondary education when the digital students begin to dominate our classrooms, schools, and state departments of education as teachers, managers, and policy makers, which is what we have started experiencing today. As indicated in many other studies (Baker 2005; Johnson et al. 2011; Schultz et al. 2008), the use of web based resources such as Web GIS, mapping systems and Google Earth have great potential to facilitate and catalyze the adoption process of GST in secondary schools. However, effective practice of these technologies will always depend upon how informed we are about their roles in our teaching activities and proper methodologies, through which their versatile benefits will be realized in our classrooms as teachers, educators, decision makers and school managers. After removing all the pedagogical and mainly teacher-centered obstacles, only then we can make sure that our efforts to avoid technological, administrative and curricular hurdles will not be in vain and will bring benefits in order to allow the use of GST in our classrooms more effectively and commonly.

Acknowledgement I would like to thank Joseph J. Kerski for his valuable contribution to this chapter with his ideas, especially about what effectiveness means in teaching and learning with GST.

References

Akerson, V. L., & Dickinson, L. E. (2003). Using GIS technology to support K-8 scientific inquiry teaching and learning. *Science Educator, 12*(1), 41–47.

Alibrandi, M. (2003). *GIS in the classroom: Using Geographic Information Systems in social studies and environmental science*. Portsmouth: Heinemann.

Audet, R. (1993). *Developing a theoretical basis for introducing Geographic Information Systems into high schools: Cognitive implications*. PhD dissertation. School of Education, Boston University, Boston.

Audet, R. H., & Abegg, G. L. (1996). Geographic Information Systems: Implications for problem solving. *Journal of Research in Science Teaching, 33*(1), 21–45.

Audet, R. H., & Paris, J. (1997). GIS implementation model for schools: Assessing the critical concerns. *Journal of Geography, 96*(6), 293–300.

Baker, T. R. (2005). Internet-based GIS mapping in support of K–12 education. *The Professional Geographer, 57*(1), 44–50.

Baker, T. R., & Bednarz, S. W. (2003). Lessons learned from reviewing research in GIS education. *Journal of Geography, 102*(6), 231–233.

Baker, T. R., & White, S. H. (2003). The effects of GIS on students' attitudes, self-efficacy, and achievement in middle school science classrooms. *Journal of Geography, 102*(6), 243–254.

Baker, T. R., Palmer, A. M., & Kerski, J. J. (2009). A national survey to examine teacher professional development and implementation of desktop GIS. *Journal of Geography, 108* (4–5), 174–185.

Bednarz, S. W. (2004). Geographic Information Systems: A tool to support geography and environmental education? *GeoJournal, 60*, 191–199.

Bednarz, S., & Audet, R. (1999). The status of GIS technology in teacher preparation programs. *Journal of Geography, 98*, 60–67.

Bednarz, S. W., & Ludwig, G. (1997). Ten things higher education needs to know about GIS in primary and secondary education. *Transactions in GIS, 2*(2), 123–133.

Bednarz, S. W., & van der Schee, J. (2006). Europe and the United States: The implementation of Geographic Information Systems in secondary education in two contexts. *Technology, Pedagogy and Education, 15*(2), 191–205.

Benimmas, A., Kerski, J., & Solís, P. (2011). The impact of a geographic technologies professional development institute on Middle Eastern and North African teachers. *International Research in Geographical and Environmental Education, 20*(1), 21–45.

Bevainis, L. (2008). Applying the GIS in school education: The experience of Japanese geography teachers. *Geografija, 44*(2), 36–40.

Biilmann, O. (2001). GIS and remote sensing in primary and secondary education: Rationale, strategies and didactics. In D. R. Green (Ed.), *GIS: A sourcebook for schools* (pp. 137–144). London: Taylor & Francis.

Breetzke, G., Eksteen, S., & Pretorius, E. (2011). Paper-based GIS: A practical answer to the implementation of GIS education into resource-poor schools in South Africa. *Journal of Geography, 110*(4), 148–157.

Chun, B. A., & Hong, I. (2007). Integrating GIS with geographic and environmental education into K-12: An interdisciplinary curriculum development entitled studying the environment of Eighteenmile Creek. *Journal of the Korean Geographical Society, 42*(2), 295–313.

DeMers, M. N., & Vincent, J. S. (2008). ArcAtlas in the classroom: Pattern identification, description, and explanation. *Journal of Geography, 106*(6), 277–284.

Demirci, A. (2008). Evaluating the implementation and effectiveness of GIS-based applications in secondary school geography lessons. *American Journal of Applied Sciences, 5*(3), 169–178.

Demirci, A. (2011). Using Geographic Information Systems (GIS) at schools without a computer laboratory. *Journal of Geography, 110*(2), 49–59.

Demirci, A., Karaburun, A., & Kılar, H. (2013a). Using Google Earth as an educational tool in secondary school geography lessons. *International Research in Geographical and Environmental Education, 22*(4), 277–290.

Demirci, A., Karaburun, A., & Ünlü, M. (2013b). Implementation and effectiveness of GIS-based projects in secondary schools. *Journal of Geography, 112*(5), 214–228.

Doering, A., & Veletsianos, G. (2008). An investigation of the use of real-time, authentic geospatial data in the K–12 classroom. *Journal of Geography, 106*(6), 217–225.

Favier, T. T., & Van der Schee, J. (2012). Exploring the characteristics of an optimal design for inquiry-based geography education with Geographic Information Systems. *Computers & Education, 58*, 666–677.

Goldstein, D., & Alibrandi, M. (2013). Integrating GIS in the middle school curriculum: Impacts on diverse students' standardized test scores. *Journal of Geography, 112*(2), 68–74.

Goodchild, M. F., & Kemp, K. K. (1990). *The NCGIA core curriculum in GIS*. Santa Barbara: National Center for Geographic Information and Analysis.

Henry, P., & Semple, H. (2012). Integrating online GIS into the K–12 curricula: Lessons from the development of a collaborative GIS in Michigan. *Journal of Geography, 111*(1), 3–14.

Huang, K. H. (2011). A GIS-interface web site: Exploratory learning for geography curriculum. *Journal of Geography, 110*(4), 158–165.

Huynh, N. T. (2009). *The role of geospatial thinking and geographic skills in effective problem solving with GIS: K-16 education*. Unpublished PhD thesis, Wilfrid Laurier University, Waterloo.

Johnson, N. D., Lang, N. P., & Zophy, K. T. (2011). Overcoming assessment problems in Google Earth-based assignments. *Journal of Geoscience Education, 59*, 99–105.

Keiper, T. A. (1999). GIS for elementary students: An inquiry into a new approach to learning geography. *Journal of Geography, 98*(2), 47–59.

Kerski, J. J. (2003). The implementation and effectiveness of geographic information systems technology and methods in secondary education. *Journal of Geography, 102*(3), 128–137.

Kerski, J. J. (2008). The role of GIS in digital earth education. *International Journal of Digital Earth, 1*(4), 326–346.

Kerski, J. J., Demirci, A., & Milson, A. J. (2013). The global landscape of GIS in secondary education. *Journal of Geography, 112*(6), 232–247.

Kidman, G. C., & Palmer, G. (2006). GIS: The technology is there but the teaching is yet to catch up. *International Research in Geographical and Environmental Education, 15*(3), 289–296.

Kim, M., Bednarz, R., & Lee, S. (2011). GIS education for teachers in South Korea: Who participates and why? *Journal of the Korean Geographical Society, 46*(3), 382–395.

Lay, J., Chen, Y., & Chi, Y. (2013). GIS adoption among senior high school geography teachers in Taiwan. *Journal of Geography, 112*(3), 120–130.

Lee, J., & Bednarz, R. (2009). Effect of GIS learning on spatial thinking. *Journal of Geography in Higher Education, 33*(2), 183–198.

Lee, J., & Bednarz, R. (2012). Components of spatial thinking: Evidence from a spatial thinking ability test. *Journal of Geography, 111*(1), 15–26.

Lemberg, D., & Stoltman, J. P. (2001). Geography teaching and the new technologies: Opportunities and challenges. *Journal of Education, 181*(3), 63–76.

Lidstone, J., & Stoltman, J. (2006). Editorial: Searching for, or creating, knowledge: The roles of Google and GIS in geographical education. *International Research in Geographical and Environmental Education, 15*(3), 205–209.

Linn, S., Kerski, J., & Wither, S. (2005). Development of evaluation tools for GIS: How does GIS affect student learning? *International Research in Geographical and Environmental Education, 14*(3), 217–224.

Liu, S., & Zhu, X. (2008). Designing a structured and interactive learning environment based on GIS for secondary geography education. *Journal of Geography, 107*(1), 12–19.

Liu, Y., Bui, E. N., Chang, C., & Lossman, H. G. (2010). PBL-GIS in secondary geography education: Does it result in higher-order learning outcomes? *Journal of Geography, 109*(4), 150–158.

Meyer, J. W., Butterick, J., Olkin, M., & Zack, G. (1999). GIS in the K-12 curriculum: A cautionary note. *Professional Geographers, 51*(4), 571–578.

Milson, A. J., & Earle, B. D. (2008). Internet-based GIS in an inductive learning environment: A case study of ninth-grade geography students. *Journal of Geography, 106*(6), 227–237.

Milson, A. J., & Kerski, J. J. (2012). Around the world with geospatial technologies. *Social Education, 76*(2), 105–108.

Milson, A. J., Demirci, A., & Kerski, J. J. (Eds.). (2012a). *International perspectives on teaching and learning with GIS in secondary schools*. New York: Springer.
Milson, A. J., Kerski, J. J., & Demirci, A. (2012b). The world at their fingertips: A new age for spatial thinking. In A. J. Milson, J. J. Kerski, & A. Demirci (Eds.), *International perspectives on teaching and learning with GIS in secondary schools* (pp. 115–137). New York: Springer.
Nellis, M. D. (1994). Technology in geographic education: Reflections and future directions. *Journal of Geography, 93*(1), 36–39.
Nielsen, C. P., Oberle, A., & Sugumaran, R. (2011). Implementing a high school level geospatial technologies and spatial thinking course. *Journal of Geography, 110*(2), 60–69.
Patterson, M. W., Reeve, K., & Page, D. (2003). Integrating Geographic Information Systems into the secondary curricula. *Journal of Geography, 102*(6), 275–281.
Perkins, N., Hazelton, E., Erickson, J., & Allan, W. (2010). Place-based education and Geographic Information Systems: Enhancing the spatial awareness of middle school students in Maine. *Journal of Geography, 109*(5), 213–218.
Roulston, S. (2013). GIS in Northern Ireland secondary schools: Mapping where we are now. *International Research in Geographical and Environmental Education, 22*(1), 41–56.
Schultz, R. B., Kerski, J. J., & Patterson, T. C. (2008). The use of virtual globes as a spatial teaching tool with suggestions for metadata standards. *Journal of Geography, 107*(1), 27–34.
Shin, E. (2006). Using Geographic Information System (GIS) to improve fourth graders' geographic content knowledge and map skills. *Journal of Geography, 105*(3), 109–120.
Shin, E. (2007). Using Geographic Information System (GIS) technology to enhance elementary students' geographic understanding. *Theory & Research in Social Education, 35*(2), 231–255.
Stoltman, J., & De Chano, L. (2003). Continuity and change in geography education: Learning and teaching. In R. Gerber (Ed.), *International handbook on geographical education* (pp. 115–137). Dordrecht: Kluwer.
Svatonova, H., & Mrazkova, K. (2010). Geoinformation technologies: New opportunities in geography education? In M. V. Zuljan & J. Vogrinc (Eds.), *Facilitating effective student learning through teacher research and innovation* (pp. 331–348). Ljubljana: University of Ljubljani.
Wang, Y. H., & Chen, C. M. (2013). GIS education in Taiwanese senior high schools: A national survey among geography teachers. *Journal of Geography, 112*(2), 75–84.
Wechsler, S. P., & Pitts, L. A. (2004). GIS in high school integrates geography with technology: A case study. *The California Geographer, 44*, 38–54.
West, B. A. (2003). Student attitudes and the impact of GIS on thinking skills and motivation. *Journal of Geography, 102*(6), 267–274.
West, B. A. (2008). *Conceptions of Geographic Information Systems (GIS) held by senior geography students in Queensland*. Unpublished PhD thesis, Queensland University of Technology, Brisbane.
Wiegand, P. (2001). Geographical Information Systems (GIS) in education. *International Research in Geographical and Environmental Education, 10*(1), 68–71.
Wiegand, P. (2003). School students' understanding of choropleth maps: Evidence from collaborative mapmaking using GIS. *Journal of Geography, 102*(6), 234–242.
Wilder, A., Brinkerhoff, J. D., & Higgins, T. M. (2003). Geographic information technologies and project-based science: A contextualized professional development approach. *Journal of Geography, 102*(6), 255–266.

Chapter 13
An Assessment of the Use of GIS in Teaching

Geok Chin Ivy Tan and Qiu Fen Jade Chen

Abstract The present research study is to provide an assessment of the use of GIS in teaching. A general evaluation of current literature was done to draw a global landscape of GIS in education. For the second part of this study, 15 research articles stemming from ten countries including USA, Germany, Northern Ireland, Ghana, Singapore, Hong Kong, Taiwan, South Korea, Australia and New Zealand written in the decade spanning across 2003–2013 were evaluated to assess the conditions encouraging and discouraging the integration of GIS usage in each country. While interest and belief in the benefits of GIS in teaching and learning among teachers have been affirmed by the research papers evaluated here, nonetheless, the adoption of GIS in schools remains low in both developed and developing countries due to a variety of factors. While common threads and challenges were identified, contextual differences among the countries present a diverse account of GIS integration into the curricula of the world today.

Keywords GIS adoption • Spatial technologies • Geography education

13.1 Introduction

Geographic Information System (GIS) is generally viewed as an excellent Information Technology (IT) tool for promoting higher-order thinking skills such as decision making and problem solving. It also provides an avenue for creating cooperative learning environments and acquiring project management skills (Palladino 1993). Hunter and Xie (2001), Kerski (1999, 2000), Meaney (2006), Rooney (1997) and several other authors also support this view and add that the authentic data used in GIS-based teaching creates the environment for real world learning, which can be both multidisciplinary and interdisciplinary, especially when used to study local issues (Jenner 2006). As Kerski (2003: 134) notes, GIS

G.C.I. Tan (✉) • Q.F.J. Chen
National Institute of Education, Nanyang Technological University, Singapore, Singapore
e-mail: ivy.tan@nie.edu.sg; jade.chen@nie.edu.sg

O. Muñiz Solari et al. (eds.), *Geospatial Technologies and Geography Education in a Changing World*, Advances in Geographical and Environmental Sciences,
DOI 10.1007/978-4-431-55519-3_13

as a teaching tool allows for an 'issues-based, student-centered, standards-based, inquiry oriented education'. His research of GIS in American high schools have led him to the conclusion that using GIS in Geography led to an increase in student motivation, allowed for new ways of communication among students and teachers, stimulated visual learners and other students who learn better through non-traditional means. Apart from a more hands-on and self-directed research experience, GIS also promotes spatial thinking involving concepts of space, visualisation and reasoning (Kidman and Palmer 2006).

While the role and use of GIS spans widely across various industries and government bodies (e.g. see Kidman and Palmer 2006), its reach into education falls short of its powerful potential to transform teaching and learning. In the long run, there is much potential for GIS to be introduced in schools especially in the teaching of subjects such as social studies, geography, history and the sciences. Kidman and Palmer (2006) have listed an astounding figure of 37 schools subjects across the areas of Natural Sciences, Social Sciences, Physical Sciences and Natural Resources that could be used with GIS capabilities in data management and spatial analysis. Even within specific subjects, the authors have found that the use of GIS in Australia appears to feature more in Geography classes than Science classes, whereas the opposite trend has been found in the USA (e.g. see Kerski 2003). Yet the reality remains that even within Geography, widely agreed by fellow researchers to be the champion of GIS, the take up rate of GIS remains low in schools across the globe (Kerski 2003; Höhnle et al. 2013; Lam et al. 2009; Wheeler et al. 2010; Yap et al. 2008) or limited to senior high school levels (Wang and Chen 2013).

While most research studies on GIS are anecdotal accounts of specific case studies, Kerski et al. (2013) have attempted a bird's eye view of 33 countries from the major world regions to examine and evaluate the status of GIS in the secondary school level across the globe. Their research has led them to acknowledge the contextual differences in the schools in the various countries, existing infrastructure, challenges, opportunities, as well as the value accorded to the role of GIS in education. While common themes and challenges can be identified in the process of GIS adoption in the various countries, they concluded that pedagogical, curricular and technological conditions are important but they do not necessarily determine the success of GIS in education.

Kerski et al. (2013) have also found that many countries have adopted GIS into the geography curriculum at the secondary school level, including Canada, UK, Australia, Germany, the Netherlands, Norway, South Africa, China and Taiwan. In Taiwan for instance, GIS was already incorporated into national curriculum for senior high (grade 12) geography from 1995. These standards were later revised in 2006 and currently, GIS is a required component in the curriculum of tenth-grade students in Taiwan (Wang and Chen 2013).

In addition, WebGIS has become increasingly important and more widely used in over desktop GIS in secondary level teaching and learning in countries including

Colombia, Denmark, Finland, France, Germany, Portugal, Spain, Norway, the Netherlands, and Japan. Apart from GIS, other spatial technologies such as global positioning system (GPS), Google Earth, virtual tours and smartphones have also been utilised and incorporated by educators into lessons. The use of GIS in the different countries also exemplify what the authors call *the three Cs* of linking students to their communities, to citizenship education, and to meaningful careers (ibid: 237). Jenner (2006) describes how students at an Australian high school are able to engage with the local community through GIS skills acquired to investigate the local issues of graffiti, vegetation, land use etc. In some cases, teachers too, engage with the community and invariably form connections with local institutions, universities and agencies (Meaney 2006).

13.2 Purpose of Study

The purpose of this paper is to provide an assessment of the use of GIS and the factors that encourage and discourage the adoption and implementation of GIS in schools. Fifteen papers stemming from ten countries across the continents were identified for the purpose of the assessment and comparison, including studies in USA (Baker et al. 2009; Kerski 2003), Northern Ireland (Roulston 2013), Germany (Höhnle et al. 2013), Ghana (Oppong and Ofori-Amoah 2011), Hong Kong (Lam et al. 2009), Taiwan (Wang and Chen 2013), Singapore (Yap et al. 2008), South Korea (Kim and Lee 2011), Australia (Kidman and Palmer 2006; Jenner 2006; Meaney 2006; Wheeler et al. 2010) and New Zealand (Chalmers 2006, West 2006). These papers were reports of GIS research studies done recently in the decade of 2003–2013 and were selected as the respective authors identified and evaluated the different strategies adopted to promote the use of GIS as well as the areas of constraints and challenges in their countries. An examination of the current literature on GIS yielded an observation: most research studies tend to focus on the challenges of GIS adoption, while few of them present a glowing report of GIS integration into education. As schools around the world grapple with the intricacies of weaving GIS into formal curriculum, it seems that for a variety of different factors, opportunities and circumstances, some countries are more successful than others in this process. In addition, both developed and developing countries can encounter the very same challenges that impede GIS usage. The aim of this paper is to firstly report cases of successful implementation of GIS in schools and highlight the favourable factors encouraging the integration of GIS in classroom teaching within these 15 studies. Secondly, the paper will evaluate the discouraging factors and challenges raised by the various authors, both in the developing and developed countries.

13.3 Encouraging Factors Leading to Successful Implementations of GIS in Schools

Studies done by many other researchers have reiterated that the successful implementation of GIS in an educational setting requires an integrated and multi-tiered response ranging from teachers, and schools, to larger, external parties from the industry and government. While many studies have confirmed the potential benefits students, teachers and the community accrue from learning and tapping into GIS technologies, it is an undeniable fact that the work involved in facilitating the penetration of GIS into everyday curriculum is often difficult and challenging. This is especially so when the responsibility of bringing GIS into lessons often fall squarely on the shoulders of geography teachers. Yet while geography teachers surveyed in many studies have expressed interest in using GIS in lessons, the reality remains that adoption of GIS in schools remains low around the world. This is regrettable, especially when, as Wheeler et al. (2010: 168) observes, GIS technology is an opportunity to elevate Geography as a subject through improving its 'vocational and academic relevance and rigour.'

Yet despite the daunting challenges, geography educators are taking up the call to establish GIS lessons in the classrooms. Various studies from the different countries selected in this report document the help received ranging from the ground level support from fellow colleagues, fellow educators in the same country, to a larger and broader scale support from the national government bodies in varying amounts that have helped in the implementation of GIS in schools. Some, like funding for basic infrastructure such as computer labs and software programmes by the government are crucial and indispensable when dealing with a technological tool. We see also how in some cases, the implementation of GIS and integration into official curriculum was largely enabled through a unified tripartite partnership of the government, industry and schools to effect changes across the educational landscape. Other journeys towards GIS implementation however, are still largely driven by teacher commitment and individual schools (Wheeler et al. 2010).

In Taiwan, Wang and Chen (2013) have identified the three distinctive strategies that were used to boost the adoption of GIS in schools: firstly, was the integration of GIS into national curriculum, which consequently led to an increase in pre-service and in-service GIS training among teachers. The second strategy was the seed school project, which saw some senior high schools in Taiwan selected to promote GIS. Seed schools were also provided with a GIS computer laboratory fully equipped with computers, software, GIS tutorials, training programs as well as teaching materials obtained from outside vendors. These schools also organized and functioned as workshop grounds for GIS training for Taiwanese teachers. Lastly, national GIS and geography competitions were often organized and served as important avenues for the promotion and usage of GIS among teachers and students.

Over in Hong Kong, Lam et al. (2009: 72) observed that aided by the financial support of the government, schools were increasingly able to afford GIS software and also that 'the availability of hardware is no longer a critical barrier to the implementation of GIS in school geography.'

Regarding software preparedness, the provision of relevant and usable digital data (which can be costly and difficult to source) was also a significant factor in bolstering the usage of GIS. In Singapore, while schools were generally well-equipped with computers and computer labs, the problem of a lack of resource packages for GIS lessons was largely resolved by the joint efforts by the Ministry of Education and the National Institute of Education in producing EduGIS, a GIS resource package that comes equipped with GIS data sets in a CD format for easy usage. Training workshops were organized to familiarize teachers with the GIS software ArcView or ArcExplorer. To further encourage the use of GIS in local schools, knowing that the cost of GIS software could be costly, MOE provided schools with a S$2000 subsidy to offset the cost of the GIS software.

Over in New Zealand, Chalmers (2006) acknowledged the efforts made not only by the government as in the case of Singapore, but also the collaboration between the government and the private sector in promoting the use of GIS through various ways that include sponsorship for GIS competitions and providing GIS data for school projects. Major software developers of GIS have also supported the use of GIS by offering their software at extremely low prices.

While hardware and software preparedness is a pre-requisite in facilitating GIS use and learning among students, the inclusion of GIS into national curriculum and examinations gives it an important weightage and validation of its place in Geography. As noted by Roulston (2013), for the first time from 2009 in Northern Ireland, GIS was specified as a requirement for students in the GCSE (14–16 years old) examinations, albeit under only one specific section. He concludes that this inclusion marks a small but significant shift and may possibly reflect the wider structural and online support for GIS available in schools.

Kidman and Palmer (2006) listed four interdependent factors that contributed to the success of a GIS program implemented in a grade 8 geography classroom in Brisbane, Australia. Firstly, teachers' commitment in finding new ways to engage students through the use of a new tool as well as continual professional skills upgrading. Secondly, the support of the school, especially that of the principal, as well as colleagues was also crucial. The partnership and support of local tertiary institutions involved in pre-service teacher training programs were also integrative in ensuring adequate exposure and training in GIS for new teachers. Lastly, the government and industry also play important roles in promoting and enhancing the use of GIS through partnerships directly and indirectly with schools.

In his discussion of the success attained in promoting the use of GIS in a high school also in Australia, Jenner (2006) elaborated on a total environment of factors and initiatives that contributed to its success. This included the combined efforts made by teachers, departments, school administration team, government bodies and the GIS industry. Arrangements were initiated and made by the department's Subject Master for teachers to attend GIS training. Consequently, professional development for teachers was further supplemented by workshops and conferences. These initiatives were supported and at times funded by the school. When teachers met with obstacles in finding their own GIS data applicable to the local Australian context, the local council and state government offered assistance. ESRI, the company that produced the GIS software *ArcView*, complemented these efforts by appointing GIS co-ordinators in schools.

When the problem of inadequate computers for GIS lessons came up, teachers from the Geography department were also, cleverly, encouraged to join the Learning Technology committee. This committee was tasked with making suggestions on how the school's resources were to be allocated when it came to IT purchases. GIS professionals were also invited to promote the uses of GIS to other teachers as well as the administration team in the school. Eventually the computers required for the implementation of GIS were acquired.

The use of GIS is not limited only to the Geography department. The Social Sciences and Science departments in the school also actively make use of GIS in their various projects. Jenner then makes the observation that such a cross-curriculum approach facilitates long term sustainability of GIS in school, in that it is not solely championed and used only by the Geography department.

And for all the efforts the teachers have made in exposing their students to the benefits of GIS, Jenner (2006) reported that the students have in turn, become competent users of GIS and have presented their research findings at local and regional settings. Indeed, at its best, GIS not only benefits students' learning and engagement with the world, the information gleaned from GIS technologies also benefits the local community.

The case studies discussed above highlight the different, and often resourceful and innovative means that have been undertaken by fellow geography educators to achieve a similar end. In the goal of increasing GIS usage and adoption, we are presented with a situation where efforts from both the top end (government, industry) and on the ground level (teachers, school) intermingle on different levels to induce varying results in the countries surveyed. The successful case studies elucidate the ideal of an effective tripartite relationship between the government, industry and schools, with teachers assuming the important role as gatekeepers and champions, in the collective efforts to integrate GIS firmly into the education systems around the world.

13.4 Discouraging Factors or Constraints to GIS Implementation in Schools

The second aim of this paper is to evaluate the discouraging factors and constraints to GIS implementation in schools as raised by the various authors. Table 13.1 provides the list all the keys factors that discouraged GIS use and adoption by teachers and schools which were synthesized from the 15 research studies. Ghana, a developing country, struggles with the lack of computers and computer labs, but so do Germany, Singapore and USA, albeit of a different nature. In Ghana, the physical infrastructure barely exists, whereas the other countries struggle with opportunities and time to use existing infrastructure during curriculum time. Even within the same country, as seen from the case studies of Australia, New Zealand and USA from the table, a problem can be experienced in one state and not the other. Hence, while diversity among the countries where the case studies originated from would be favourable in our report, looking into several case studies from a country can yield various insights too. However, in general, one can see that the key factors discouraging the use of GIS from these 15 research studies pertain to the hardware and software issues, the need or lack of GIS training for teachers, the lack of time and the nature of the curriculum.

13.4.1 Hardware

Teachers in Hong Kong voiced their concerns that they did not have ready access to the computers or had to compete with the other subjects and teachers for the use of special computer rooms (Lam et al. 2009). Studies in America (see Baker et al. 2009) and Northern Ireland (Roulston 2013) have found that not all schools have sufficient computers to enable all students to have access to GIS programmes. In Ghana, in particular, the hardware is a major setback (Oppong and Ofori-Amoah 2011). The authors stated that there was even a lack of uninterrupted access to electricity in the rural areas. In the capital city, frequent interruption in electricity supply was also common and that would disrupt the effective use of computers in schools.

13.4.2 Software

Besides hardware problems, several software issues were identified by 11 out of the 15 research studies reviewed here. These issues were the cost of the software; the complexity of the software; the lack of suitable GIS resources and teaching materials; the lack of digital data and need for curriculum support. Lam et al. (2009) stated that there were schools in Hong Kong without financial support to acquire GIS software and schools which acquired outdated versions of ArcView needed financial support for upgrading. Another major problem the authors raised was that digital data was expensive. In most countries surveyed in the research studies reviewed here, sometimes the use of software itself can be challenging to both teachers and students. Addressing both hardware and software problems is especially pertinent and the first step in ensuring a successful implementation of GIS. As pointed out by Jenner (2006) in his experience as a high school teacher in Australia, these issues had to be ironed out in the early stages of planning, as

Table 13.1 Factors discouraging the use of GIS in the 15 research studies

Factors	USA (Kerski 2003)	USA (Baker et al. 2009)	Germany (Höhnle et al. 2013)	North Ireland (Roulston 2013)	Ghana (Oppong and Ofori-Amoah 2011)	Singapore (Yap et al. 2008)	Hong Kong (Lam et al. 2009)
Need/lack GIS training	✔		✔	✔	✔	✔	✔
Software: High cost/lack GIS resources/digital data	✔		✔	✔	✔	✔	✔
Hardware: Lack computer/computer lab		✔	✔	✔	✔	✔	✔
Teachers lack time	✔	✔	✔	✔		✔	
Teachers lack GIS knowledge			✔				✔
Teachers lack confidence in GIS							✔
Lack GIS expertise/support/practice		✔	✔				
Few teachers trained in GIS							
Low/debatable merits of GIS			✔				
High teaching workload			✔				
Students lack GIS skills		✔					
GIS is optional/not part of curriculum					✔	✔	✔
Tight curriculum/limited class hours		✔	✔		✔	✔	
No access to electricity					✔		

Taiwan (Wang and Chen 2013)	South Korea (Kim and Lee 2011)	Australia (Wheeler et al. 2010)	Australia (Meaney 2006)	Australia (Jenner 2006)	Australia (Kidman and Palmer 2006)	New Zealand (West 2006)	New Zealand (Chalmers 2006)
✓	✓	✓	✓	✓		✓	✓
✓	✓	✓	✓	✓	✓		✓
✓	✓	✓		✓		✓	
		✓		✓	✓		✓
		✓					
✓			✓		✓	✓	✓
				✓	✓	✓	
✓						✓	
						✓	
✓		✓			✓		

teachers had to commit time (often after school hours) to learn the GIS software, as well as possibly adjust their teaching styles in accordance to the unique challenges and demands of the teaching and learning of GIS for students.

The lack of available GIS teaching units focusing on the local issues while keeping within the school syllabus was commonly raised as another issue. As a result, much time is needed to develop and modify units to focus on local issues and relevant topics. To develop and modify the resources, the teachers will run into difficulties in obtaining the much needed digital data. With few GIS integrated teaching units, it becomes a viscous cycle. In Germany, Höhnle et al. (2013) found that teachers and students lose the skills and confidence in using GIS when they lack regular practice. The lack of simple and copyable GIS lessons is seen as a major impediment. The German geography teachers craved for simple small examples, which can be easily copied without much technical know-how and can be transferred to their local surroundings. This is a hardly a surprising request and is frequently echoed by geography teachers around the world.

13.4.3 Teacher Training

GIS training for teachers has been identified as essential and the lack of it is another major impediment. There is a need for specialised GIS skill training to support classroom use of GIS (Roulston 2013). West (2006) also stated that teacher training should shift the focus from teaching what GIS is to how to teach with GIS. Even after receiving the training, additional support for teachers is also needed for them to become competent and confident in using GIS.

13.4.4 Teacher's Time

The lack of time is another salient challenge for the teachers trying to use GIS in their teaching. In his assessment of GIS use in the New Zealand context, Chalmers (2006) stated that his biggest concern was not with software or hardware but teachers' time. On the same note, teachers also lack time to develop GIS-based lessons as was reported in Wheeler et al. (2010), Baker et al. (2009) and Kerski (2003). Teachers are hard pressed for time as a result of very tight curriculum and limited hours to cover the syllabus. The use of GIS is not seen as effective when teachers are trying hard to cover the tight curriculum. Time is also needed for teacher to gain confidence in using GIS and to attend GIS professional development and workshops (Kidman and Palmer 2006). According to Höhnle et al. (2013), German teachers hardly use GIS tools at home and it would mean that when they want to use it in class they would have to repetitively practise and get to know how to use GIS tools all over again after a period of not using it. Hence, the use of GIS has been deemed as far too time consuming by teachers, especially after considering how much work is required in the preparation for the few geography lessons to be taught.

13.4.5 Tight Curriculum

The curriculum was another issue raised in some of the research papers reviewed here. In Australia for example, geography is not taught as a separate subject but as a small part of humanities (Wheeler et al. 2010). The demise of geography and the move toward integrated studies have led to a reduced number of teachers trained in the teaching of geography in Australia and New Zealand (West 2006). As for Hong Kong, the use of GIS in classrooms was perceived to be impractical and the use of GIS was not mandatory in the current secondary geography curriculum (Lam et al. 2009). The limited class hours and limited curriculum time were key obstacles to the use of GIS in Taiwan (Wang and Chen 2013). Many of the other research papers in America, Europe and Asia-Pacific regions had teachers articulating that there was just no time to introduce GIS into the already overcrowded curriculum (e.g. Kidman and Palmer 2006; Lam et al. 2009; Baker et al. 2009).

Having examined the factors that promote and impede the successful implementation of GIS, we can perhaps come to the conclusion that the efforts of the geography teachers are not enough. GIS is a technological tool that requires pre-existing infrastructure and equipment, digital data, time, knowledge and skills on the part of the user before it yields its powerful capabilities. These are further bound in a complex web that involves administrative, structural and ground level realities that need to be in sync, which is difficult to say the least. As we have seen from the various scenarios described in different countries, there is often a lack of an integrated and coordinated national policy and support. The cost to set up GIS is often high. This is further coupled with a curriculum structure that leaves little time on the part of teachers who already feel unconfident, unsupported and do not always see the worth of investing time and effort into GIS when tangible benefits are not always seen. And to state the obvious, not all Geography teachers are proponents of GIS. It is unsurprising hence, to see that GIS adoption and use is still few and far and successful only in a few selected cases.

13.5 Conclusion

The research findings indicated that teachers in many countries were still not encouraged to use GIS in the teaching of geography because of various factors. The key common constraining factors as derived from 15 recent research papers from 10 nations include accessibility to hardware; complexities of software and lack of GIS-based resources; insufficient or lack of GIS teacher training; lack of time; and the tight curriculum.

It is also noted that most of the GIS-based lessons were conducted to cover the existing syllabus using prepared GIS resource packages or modified GIS resource packages. Teachers are highly dependent on these pre-prepared GIS resource packages and provision of relevant and digital data. In Singapore too, the Ministry of Education and the National Institute of Education were also concerned about developing GIS resource packages to enable teachers to use in the classroom (Zhu et al. 2006). Perhaps, there should be a turning point away from providing teachers with ready-made GIS resource packages and to move towards using GIS not just as a technological tool but as a mode of instruction which requires students to collect authentic spatial data to solve real problems. The students should thereby be able to construct knowledge through mapping and analysing their own data to explore spatial patterns and relationships.

The real challenge, then perhaps, is the need for a paradigm shift in the goal and expectations of both teachers and students (Kerski 1999, 2000, 2003) and not just to understand the factors that can impede the implementation of GIS. Integrating GIS into classroom learning is a complex process (Kerski 2000, 2003). Successful integrating of GIS should be facilitated by a total reorganization of the curriculum and a shift towards problem-based learning (Bednarz 2001). More importantly, teachers need more than just practice in the use of GIS. They must buy into the idea that GIS can engage students to make their own observation of patterns on the maps and even help with problem solving.

References

Baker, T. R., Palmer, A. M., & Kerski, J. J. (2009). A national survey to examine teacher professional development and implementation of desktop GIS. *Journal of Geography, 108* (4-5), 174–185.

Bednarz, S. W. (2001). Thinking spatially: Incorporating geographic information science in pre and postsecondary education. In L. T. Houtsonen & M. Tammilehto (Eds.), *Innovative practices in geographical education: Proceedings of the Helsinki Symposium of the International Geographical Union, Commission on Geographical Education* (pp. 3–7). Finland: University of Helsinki.

Chalmers, L. (2006). GIS in New Zealand schools: Issues and prospects. *International Research in Geographical and Environmental Education, 15*(3), 268–270.

Höhnle, S., Schubert, J. C., & Uphues, R. (2013). What are the constraints to GIS usage? Selected results of a teacher survey about constraints in the school context. *International Research in Geographical and Environmental Education, 22*(3), 226–240.

Hunter, B., & Xie, Y. (2001). Data tools for real-world learning. *Learning & Leading with Technology, 28*(7), 18–24.

Jenner, P. (2006). Engaging students through the use of GIS at Pimlico State High School. *International Research in Geographical and Environmental Education, 15*(3), 278–282.

Kerski, J. J. (1999). *A nationwide analysis of the implementation of GIS in high school education.* http://gis.esri.com/library/userconf/proc99/proceed/papers/pap202/p202.htm. Accessed 7 May 2014.

Kerski, J. J. (2000). *The implementation and effectiveness of Geographic Information Systems Technology and methods in secondary education.* Unpublished doctoral dissertation, University of Colorado, Boulder, USA.

Kerski, J. J. (2003). The implementation and effectiveness of Geographic Information Systems technology and methods in secondary education. *Journal of Geography, 102*(3), 128–137.

Kerski, J. J., Demirci, A., & Milson, A. J. (2013). The global landscape of GIS in secondary education. *Journal of Geography, 112*(6), 232–247.

Kidman, G., & Palmer, G. (2006). GIS: The technology is there but the teaching is yet to catch up. *International Research in Geographical and Environmental Education, 15*(3), 289–296.

Kim, M., & Lee, S. I. (2011). South Korea: GIS implementation profiles among secondary geography teachers. In A. J. Milson, A. Demirci, & J. J. Kerski (Eds.), *International perspectives on teaching and learning with GIS in secondary schools* (pp. 233–241). New York: Springer.

Lam, C. C., Lai, E., & Wong, J. (2009). Implementation of geographic information system (GIS) in secondary geography curriculum in Hong Kong: Current situations and future directions. *International Research in Geographical and Environmental Education, 18*(1), 57–74.

Meaney, M. (2006). Towards Geographic Information Systems (GIS) implementation: A case study. *International Research in Geographical and Environmental Education, 15*(3), 283–288.

Oppong, J. R., & Ofori-Amoah, B. (2011). Ghana: Prospects for secondary school GIS education in a developing country. In A. J. Milson, A. Demirci, & J. J. Kerski (Eds.), *International perspectives on teaching and learning with GIS in secondary schools* (pp. 115–123). New York: Springer.

Palladino, S. D. (1993). *GIS in the Schools: Workshop Resource Packet.* National Center for Geographic Information and Analysis Technical Report 93-2. http://www.ncgia.ucsb.edu/Publications/Tech_Reports/93/93-2.PDF. Accessed 8 May 2014.

Rooney, P. E. (1997). A spatial tool for schools. *Hands On! 20*(1), 4–5.

Roulston, S. (2013). GIS in Northern Ireland secondary schools: Mapping where we are now. *International Research in Geographical and Environmental Education, 22*(1), 41–56.

Wang, Y. H., & Chen, C. M. (2013). GIS education in Taiwanese senior high schools: A national survey among geography teachers. *Journal of Geography, 112*(2), 75–84.

West, B. A. (2006). Forum: Issues affecting the adoption of GIS within Australian and New Zealand schools. *International Research in Geographical and Environmental Education, 15*(3), 255–258.

Wheeler, P., Gordon-Brown, L., Peterson, J., & Ward, M. (2010). Geographical information systems in Victorian secondary schools: Current constraints and opportunities. *International Research in Geographical and Environmental Education, 19*(2), 155–170.

Yap, L. Y., Tan, G. C. I., Zhu, X., & Wettasinghe, M. C. (2008). As assessment of the use of Geographical Information Systems (GIS) in teaching geography in Singapore schools. *Journal of Geography, 107*, 52–60.

Zhu, X., Liu, Y., Yap, L. Y., & Tan, G. C. I. (2006). Issues in the design of GIS resources for secondary education in Singapore. In K. C. Goh & Y. Sekson (Eds.), *Change and development in Southeast Asia in an era of globalisation* (pp. 419–428). Singapore: Prentice Hall.

Chapter 14
A Review of Geospatial Thinking Assessment in High Schools

Bob Sharpe and Niem Tu Huynh

Abstract The types of assessments found in geography teaching materials generally focus on declarative knowledge, include few geographic practices, are primarily multiple-choice format, and use a map/globe as the type of representation or none at all. With this context in mind, results are presented from an online survey of geospatial thinking assessment practices of educators involved in high school geography education from around the globe. The findings lead to a discussion of the assessment practices teachers use to measure high school students' performance in geospatial thinking, the concepts they most often assess and the challenges they encounter in undertaking such assessments. The chapter concludes by recommending the development of new online tools for geospatial thinking assessment, and urging teachers to put into practice the ideas and instruments being proposed in the emerging literature on the assessment of geospatial thinking.

Keywords Assessment practices • Geospatial thinking • Geography teachers • High school

14.1 Introduction

The chapter focuses on advances in the literature regarding the assessment of geospatial thinking, and how geography teachers are putting these ideas into practice. The chapter first reviews the scholarly journals to identify recent research trends in the assessment of geospatial thinking in the high school context. The second part describes the method used to gather information on a cross section of assessment practices used by high school geography teachers from around the globe. Using an online survey, educators involved in high school geography

B. Sharpe (✉)
Wilfrid Laurier University, Waterloo, ON, Canada
e-mail: bsharpe@wlu.ca

N.T. Huynh
Association of American Geographers, Washington, DC, USA
e-mail: nhuynh@aag.org

O. Muñiz Solari et al. (eds.), *Geospatial Technologies and Geography Education in a Changing World*, Advances in Geographical and Environmental Sciences,
DOI 10.1007/978-4-431-55519-3_14

education were invited to comment on geospatial thinking assessment practices in their local school districts, as well as to elaborate on their own teaching practices. Findings from the survey are then summarized in the third part of the chapter to identify common and notable practices, as well as to identify key challenges and ways forward in the assessment of geospatial thinking in high school geography classrooms.

14.2 Literature

Assessments primarily serve the purpose of evaluating learners' progression understanding a topic, skill, or developing motivation, attitude, perspective etc. This may take the form of a singular or a compilation of multiple tasks. These may be applied in a formal (e.g., classroom) or informal (e.g., summer camp) setting. Assessments are important not only because they indicate student learning (or lack thereof), but these findings are informative to teachers, administrators, and policy makers (Edelson et al. 2013).

Edelson et al. (2013) argue that assessment informs learning and teaching, and vice versa. It may play a role in teachers' instructional decision making (e.g., teachers' use of formative assessment to indicate difficult topics that require review), individual achievement (e.g., a mid-term exam that serves as a summative assessment of student learning), program evaluation (e.g., professional development for teachers whose students performed poorly), and education research (e.g., assessments used to validate learning progressions research).

Published assessments on student learning in geography are diverse in purpose and form. Among these, and perhaps the best known, are large-scale assessments of geography and geographic literacy commonly conducted at the state or national levels. For example, in the U.S., the National Assessment of Educational Progress (NAEP) measures student learning at grades 4, 8, and 12, of which geography is one of the subjects tested. At the secondary school level, the Advanced Placement (U.S.) and International Baccalaureate (International) are offered annually at select schools. Also familiar are large-scale literacy tests created to pique public interest such as the 2006 National Geographic Roper survey,[1] and the 2005 geographic literacy survey undertaken by Royal Canadian Geographic Society (RCGS),[2] as well as annual student competitions,[3] and the daily quizzes by National Geographic.[4]

[1] http://www.nationalgeographic.com/roper2006/index.html

[2] http://www.cgeducation.ca/programs/geoliteracy/

[3] http://www.geochallenge.ca/geochallenge/challenge.asp

[4] http://www.nationalgeographic.com/geobee/quiz/today/

More relevant to the discussion in this chapter is the scholarship on the assessment of geographic learning in the classroom. The geography literature provides evidence of assessment in the form of tasks, questions, and surveys, designed to test different facets of geography understanding: factual world knowledge (Bein et al. 2009; Cross 1987; Dunn 2011; Turner and Leydon 2012); spatial concepts such as projection (Olson 2006; Battersby and Montello 2009; Battersby and Kessler 2012) and map overlay (Albert and Golledge 1999; Battersby et al. 2006; Lee 2005; Lee and Bednarz 2009); spatial tasks (Golledge et al. 2008; Lee 2005); sense of direction and wayfinding (Montello et al. 1999; Hegarty et al. 2002); and map reading (Lanca and Kirby 1995; Liben and Yekel 1996; Kastens and Liben 2007). Map quizzes have been popular as assessments of location knowledge in post-secondary education. These vary in format and include open-ended responses, multiple choice questions, listing, labelling, and sketching, served in both online and paper forms. In addition to research-based publications of geospatial questions, textbooks provide another source offering a range of questions that vary in complexity (Jo et al. 2010; Jo and Bednarz 2011; Scholz et al. 2014).

More specific instances of assessment use in physical geography, primarily reported at the post-secondary level, include reflective diaries (e.g., Dummer et al. 2008; McGuinness 2009), which move students beyond simple data and knowledge acquisition to focus on observing and recording information. Writing and reflection encourage students to think about the research processes associated with wider theories and concepts (Dummer et al. 2008). Examples can be found of problem-solving through the application of geospatial tools in fieldwork settings (Hupy 2011). Physical geography has also assessed student abilities to communicate findings through such means as the development of websites (e.g., France and Ribchester 2004) and podcasts (e.g., Kemp et al. 2012). Examples in human geography include the use of concept mapping (Wehry et al. 2012) and journal writing (Hooey and Bailey 2005; Warkentin 2011).

There is also a literature on the implementation of Geographic Information System (GIS) instruction in the undergraduate and K-12 classrooms (Madsen and Rump 2012; Miller et al. 2005), which examines both curriculum content and instructional strategies. Little has been published, however, on practices for assessing learning in GIS classes. The only hint of a survey is on professional development, post-training learning retention (e.g., Baker et al. 2009).

An important characterization and criticism of much of the geography assessment research is that it has been developed for singular use (i.e. in the classroom or for research), is rarely replicated, and lacks validity. This, however, is changing with a growing literature on the assessment of spatial thinking. Recent published assessments (e.g., Hegarty et al. 2002; Huynh and Sharpe 2013; Kim and Bednarz 2013; Lee and Bednarz 2009) were developed with validity and reliability in mind and have undergone multiple reiterations of revision based on pilot feedback. The survey conducted for this paper indicates that these assessments are now being used for classroom and research purposes.

14.3 Method

14.3.1 Objectives

As the literature review suggests, the published research on geospatial thinking and assessment reflects a North American focus and pays scant attention to the secondary school context. In order to enhance current research and to address its deficiencies this study set out to develop a more global understanding of the geospatial thinking assessment practices used by secondary school teachers. The method adopted is a simple online survey, as described below.

14.3.2 A Survey of Geospatial Thinking Assessment Practices, Concepts and Challenges

The main objective of the survey was to determine the types and prevalence of geospatial thinking assessment practices used by teachers. A geospatial assessment is defined as one that measures student understanding of spatial relations (Huynh and Sharpe 2013). For the purpose of this survey, these assessment practices were defined as questions, exercises and activities used to measure student understanding of concepts and skills as applied to spatial reasoning within a geography context.

The survey was also intended to identify the main challenges faced by educators in the implementation of geospatial assessment practices. Evidence was sought of the challenges previously identified throughout the literature including: the lack of awareness of available geospatial assessment instruments; the lack of access to technologies and resources; and the lack of teacher training in the practice of geospatial thinking assessment. Survey respondents were asked if these or other obstacles posed a challenge to the implementation of geospatial assessment.

Drawing on the literature, the survey included questions designed to find evidence of the range of practices used for assessing geospatial thinking including: questions in geography-related textbooks; in-class multiple choice quizzes; in-class map, atlas, globe reading exercises; online virtual-globe reading exercises; in-lab problem-solving using GIS; field-based problem-solving using Global Positioning System (GPS); and assessment practices outside formal learning environments. Respondents to the survey were asked directly about the prevalence of these practices in their school district and in their own teaching.

Assessment practices are typically applied to evaluate a student's understanding and ability to apply a number of core concepts in geospatial thinking. Although the literature offers numerous typologies of spatial and geographic concepts, there has not yet been a consistent set of terms widely adopted by educators. A typology of spatial concepts was adopted for this survey that was drawn from the work published online by the Center for Spatial Studies at the University of California. The rationale for using these eight concepts is, in part, that they focus on spatial reasoning in the use of geographical information. Further, "They can be rendered understandable through simple illustrations to young children but they are also sufficiently engaging at advanced levels for thinking about scientific and social problems."[5] The concepts include: location; distance; network; neighbourhood and region; scale; spatial heterogeneity; spatial dependence; and, objects and fields.

14.3.3 Profile of Survey Respondents

The survey was distributed electronically to 43 individual educators in the authors' network and seven listservs including: International Network for Learning and Teaching Geography in Higher Education (INLT); GeoEd, GISEd, GIS Research group (gisEd@googlegroups.com), AP Human Geography group, IGU Commission for Geographic Information group, and the T3G Institute (Esri). The survey was made available online for approximately 6 weeks in early 2014. In total there were 44 usable responses collected from 15 different countries: 19 from the USA, eight from Canada, two each from Germany, the Netherlands and the UK, as well as single responses from Australia, Austria, Brazil, China, Ecuador, Indonesia, Nigeria, Portugal, Rwanda and Vietnam. Given the number of responses from the USA and Canada, in combination with the working local knowledge of the authors, the research findings for these countries can be considered authoritative. Responses from educators in other countries also provide a useful synoptic snapshot of geospatial thinking assessment practices used elsewhere around the globe.

Half of the respondents were teachers at secondary schools. The definition of secondary or high school does vary to a small extent among school districts, but in most cases these labels correspond to the equivalent of grades 9 through 12 in the North American context. Other respondents included educators affiliated with universities or colleges (18 of the 44). The large majority of respondents were affiliated with public institutions, although four represented private schools or other organizations. A few respondents indicated that they taught outside a formal learning environment, typically in a summer camp setting.

Among the individuals responding, there was a wide range of teaching experience, with one respondent teaching for 2 years and another for over 40 years. It is interesting to note that most respondents were experienced teachers reporting that they had been teaching for 16 years or more (see Fig. 14.1).

In general, the online survey proved to be an effective method for soliciting a mix of closed-ended and open-ended responses. Although the majority of responses were brief, a few respondents provided detailed comments and references, while a few attached documents with examples of their assessment instruments.

[5] http://spatial.ucsb.edu/

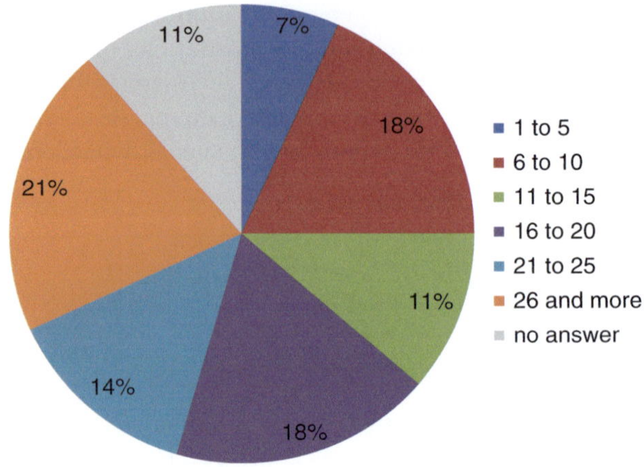

Fig. 14.1 Years of teaching experience, n = 44

14.4 Findings

14.4.1 Assessment Practices

Those responding to this survey were self-selected and presumably more likely to be educators who incorporate geospatial thinking into their teaching. It is of little surprise therefore that the majority of respondents (63 %) indicated that they used some type of geospatial thinking assessment practice. Furthermore, most teachers commented on the importance of geospatial thinking assessment and their desire to see more resources and training in these practices.

Several respondents observed that at the scale of their school district, geospatial assessments were not used, rarely used or unevenly used. In one instance a teacher noted the separation of the geography curriculum between high school and university such that geospatial concepts and skills are taught in the university but not in the schools. Teachers from several districts noted that in their districts geography courses are not required, let alone geospatial technologies (GSTs) or spatial thinking. Another respondent stated that "…there is little conceptual thinking incorporated in geography education in this country. Geography classes are usually based on everyday knowledge with little reference to the academic discipline." Still another observed that although there are some references in the curriculum to geospatial assessments, it is neither explicit, nor subject to assessment.

In contrast, school districts in Ontario and Colorado have made geospatial concepts and skills much more explicit. For example, Ontario's recently revised curriculum includes a continuum of map, globe and graphing skills (Ontario Ministry of Education 2013).[6] The continuum outlines in detail the skills to be achieved by students as they progress from grades 1 through 12. Likewise,

[6] http://www.edu.gov.on.ca/eng/curriculum/secondary/canworld910curr2013.pdf

Colorado's revised academic standards in social studies specify a series of geospatial skills and concepts or 'graduate competencies' to be mastered from preschool through twelfth grade (Colorado Department of Education).[7] In neither instance, however, do these districts adopt standardized assessment instruments or practices.

The online survey results suggest that the use of standardized assessment instruments is not widespread. A majority (63 %) of respondents indicated that they were not aware of any particular assessment instrument being used. As previously noted in the literature review, there are a few published assessment instruments that have been tested for validity and reliability which are now available for use in classroom (e.g., Hegarty et al. 2002; Huynh and Sharpe 2013; Kim and Bednarz 2013; Lee and Bednarz 2009). However, only four respondents indicated that they were using one of these standardized assessment instruments. The instruments mentioned include the *spatial thinking ability test* (Lee and Bednarz 2009), and a scale for measuring *geospatial thinking expertise* (Huynh and Sharpe 2013). These assessment instruments are being used for research purposes and curriculum assessments rather than to support classroom instruction practices.

More characteristic of teaching practices are classroom and desktop assessments including map, atlas and globe reading exercises, as well as multiple-choice quizzes (see Fig. 14.2). In most cases these take the form of paper-based tests of location knowledge and place-naming. Answering questions provided by geography-related textbooks is the next most common classroom activity. Although much less commonly reported, a similar form of testing uses online activities including virtual globes and online GIS. In general, these forms of assessment tend to emphasize student learning of declarative geospatial knowledge rather than configurational and procedural knowledge. Furthermore, as desktop activities, they also tend to limit the scale of the assessment to the classroom or virtual space rather than incorporate large geographic scales.

The frequency of teacher responses across these assessment practices was further analyzed to determine if there were any notable variations among jurisdictions. For example, it was hypothesized that in-lab problem-solving using GIS might be more prevalent in North American teaching jurisdictions. This analysis, however, did not suggest any systematic differences in assessment practices among teachers. This finding, however, may be a function of the small sample size, and is worthy of further investigation.

Although the assessment instruments used by some teachers are from published sources, over 50 % of the teachers indicated that they devised their own assessment instruments. Notably, several teachers reported that they made extensive use of field-based activities including sketching, sketch mapping, journal writing and taking field notes. One teacher reported their frequent use of mobile devices such as smartphones and tablets to have students collect, organize and present their

[7] http://www.cde.state.co.us/cosocialstudies/statestandards

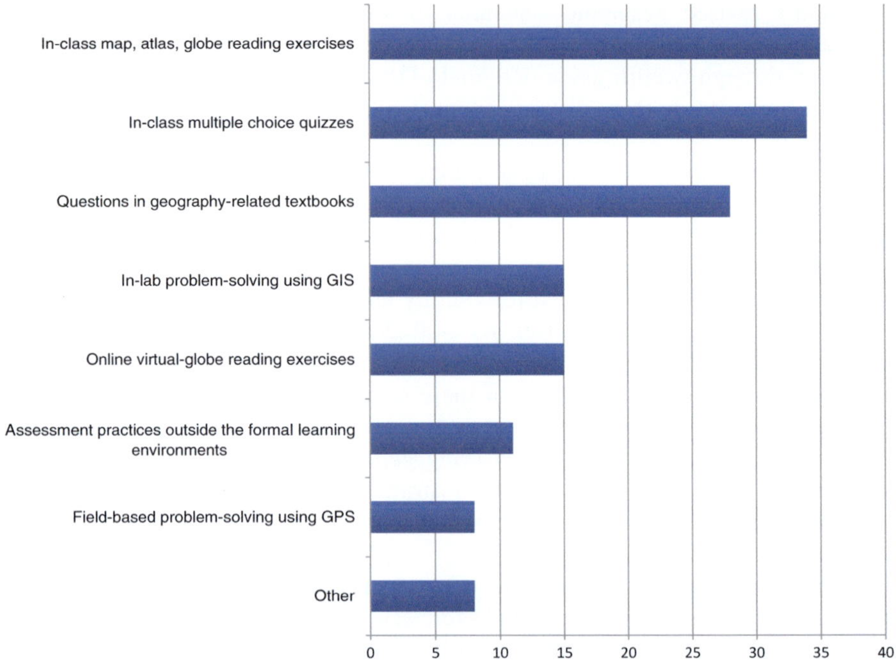

Fig. 14.2 Frequency of assessment practices commonly used by teachers, n = 44

fieldwork. Another respondent provided a link to website that has been developed to apply geospatial skills in environmental problem-solving.[8] Overall, however, the more sophisticated, inquiry-based techniques such as problem-solving activities using GIS, and field-based problem solving using GPS were the types of assessment practice reported least often.

14.4.2 Assessment of Geospatial Concepts

The geospatial concept assessed most often by teachers is *location*. All of the respondents who reported undertaking assessments incorporated the concept of location into their activities. Almost as important, were activities that assessed understanding of *distance*, *scale*, *neighbourhood and region*. The concepts least often reported included networks, spatial heterogeneity, spatial dependence and objects and fields. When asked what concepts were missing from this list, the responses suggested that teachers were not entirely familiar with the particular terminology adopted by the survey. Several respondents commented that these

[8] http://www.ei.lehigh.edu/eli/index.html

terms were primarily spatial concepts and failed to capture the geospatial or geographic and environmental content that is the emphasis of their own teaching. For example, concepts important to geography teaching and yet not explicitly mentioned on the list, include place, direction, spatial pattern and diffusion. In addition, some teachers suggested that the survey tended to overemphasize concepts rather than the application and skills in map reading, geographical thinking, environmental interpretation and the framing of geographical questions.

14.4.3 Challenges to Geospatial Assessment

The main challenge to geospatial assessment, as reported by 84 % of respondents, is the lack of teacher training in such practices. A smaller proportion (50 %) reported a lack of awareness of available geospatial assessment instruments. Lack of access to technologies and resources, was reported as a challenge by less than half the teachers, although this remains a significant barrier in some school districts, especially those outside of North America. A challenge, previously noted, is that in many districts geospatial thinking is not an explicit dimension of the local curricula. Another is that existing assessment instruments typically have been designed for particular geographic contexts and require customization to the local curriculum and local geography.

14.5 Discussion and Conclusions

Overall, the survey results suggest a degree of global consensus among geography educators regarding the value of core concepts in high school geography, including location, distance, scale, neighbourhood and region. Furthermore, given that respondents were from around the globe and from a range of social sciences and sciences suggests that the need for geospatial thinking assessment crosses both international and subject boundaries.

The survey findings also indicate that the application and further development of valid and reliable geospatial assessment practices have been slow and uneven. Although standardized assessment instruments are now available, there is little evidence that they have been widely applied. The challenges to assessment are less often a lack of technology, and more often a lack of awareness and training among teachers. Teachers could benefit from the publication of more examples of assessment instruments that could be used in geography classrooms for diagnostic, formative and summative assessments. It could be helpful to model when and how geospatial assessments are best used in the classroom through professional development, both in-service and pre-service. Furthermore, insights from such assessments could also inform curriculum reform.

Another challenge to the more widespread application of geospatial assessment instruments is the need for their adaptation to the local curriculum and local geography. The publication of further applications and refinements to the existing assessments would provide useful examples for educators in a variety of contexts.

In general, survey respondents report a positive trend which is that in many districts curricular reforms are placing increased emphasis on geospatial reasoning. However, this survey also suggests that there has been less attention given to the development of instruments and practices for the assessment of student learning outcomes. This offers an opportunity for action. The authors recognize that GSTs and applications provide a versatile platform for teaching and learning. Although such technologies may not always improve student achievement, they may be useful to enhance student motivation.

Looking ahead, the authors recommend that new tools be developed to assess the use of geospatial thinking practices. As suggested by the survey, the chief obstacle to the majority of respondents is their lack of awareness of the current strategies and instruments available to assess geospatial thinking. Computers have become a common way for teachers to demonstrate and develop mapping skills. This suggests the potential for online mapping platforms (e.g., ArcGIS Online), along with smartphone and tablet applications to become a means for teachers to access and further refine assessment instruments for geospatial thinking in different educational settings.

Assessment is an integral component of education, connected to teaching and learning. The types of assessments found in geography teaching materials generally focus on declarative knowledge, include few geographic practices, are primarily multiple-choice format, and use a map/globe as the type of representation or none at all. To enhance educator awareness of geospatial thinking assessment it will be vital to build capacity in the geographic education community. A recent effort in this direction is the road Map for 21st Century Geography Education Project. Each of its three reports highlights key recommendations for a different audience. Teachers who are interested in creating geospatial thinking assessments may want to examine the *Assessment* report (Edelson et al. 2013), which addresses key considerations, and provides a framework for designing geography assessments. For textbook and curriculum developers, the *Instruction materials and professional development* report (Schell et al. 2013) is rich with support materials where assessment tools may fit. Researchers will find that the third report, *Geography education research* (Bednarz et al. 2013) outlines ideas for advancing geographic education including recommendations for research in the area of assessment.

Acknowledgements The authors would like to thank the educators who responded to the survey and shared their insights and experiences. Appreciation is also extended to reviewers and editors for their thoughtful feedback.

References

Albert, W. S., & Golledge, R. (1999). The use of spatial cognitive abilities in geographical information systems: The map overlay operation. *Transactions in GIS, 3*(1), 7–21.

Baker, T. R., Palmer, A. M., & Kerski, J. J. (2009). A national survey to examine teacher professional development and implementation of desktop GIS. *Journal of Geography, 108* (4), 174–185.

Battersby, S., & Kessler, F. C. (2012). Cues for interpreting distortion in map projections. *Journal of Geography, 111*(3), 93–101.

Battersby, S., & Montello, D. R. (2009). Area estimation of world regions and the projection of the global scale cognitive map. *Annals of the Association of American Geographers, 99*(2), 273–291.

Battersby, S., Golledge, R., & Marsh, M. (2006). Incidental learning of geospatial concepts across grade levels: Map overlay. *Journal of Geography, 105*(4), 139–146.

Bednarz, S. W., Heffron, S., & Huynh, N. T. (Eds.). (2013). *A road map for 21st century geography education: Geography education research* (A report from the assessment committee of the road map for 21st century geography education project). Washington, DC: National Geographic Society.

Bein, F. L., Hayes, J. J., & Jones, T. G. (2009). Fifteen year follow-up geography skills test administered in Indiana, 1987 and 2002. *Journal of Geography, 108*(1), 30–36.

Cross, J. A. (1987). Factors associated with students' place location knowledge. *Journal of Geography, 86*(2), 59–63.

Dummer, T. J. B., Cook, I. G., Parker, S. L., Barrett, G. A., & Hull, A. P. (2008). Promoting and assessing 'Deep Learning' in geography fieldwork: An evaluation of reflective field diaries. *Journal of Geography in Higher Education, 32*(3), 459–479.

Dunn, J. M. (2011). Location knowledge: Assessment, spatial thinking, and new National Geography Standards. *Journal of Geography, 110*(2), 81–89.

Edelson, D. C., Shavelson, R. J., & Wertheim, J. A. (Eds.). (2013). *A road map for 21st century geography education: Assessment* (A report from the assessment committee of the road map for 21st century geography education project). Washington, DC: National Geographic Society.

France, D., & Ribchester, C. (2004). Producing websites for assessment: A case study from a level 1 fieldwork module. *Journal of Geography in Higher Education, 28*(1), 49–62.

Golledge, R., Marsh, M., & Battersby, S. (2008). A conceptual framework for facilitating geospatial thinking. *Annals of the Association of American Geographers, 98*(2), 285–308.

Hegarty, M., Richardson, A. E., Montello, D. R., Lovelace, K., & Subbiah, I. (2002). Development of a self-report measure of environmental spatial ability. *Intelligence, 30*(5), 425–447.

Hooey, C. A., & Bailey, T. J. (2005). Journal writing and the development of spatial skills. *Journal of Geography, 104*(6), 257–261.

Hupy, J. P. (2011). Teaching geographic concepts through fieldwork and competition. *Journal of Geography, 110*(3), 131–135.

Huynh, N. T., & Sharpe, B. (2013). An assessment instrument to measure geospatial thinking expertise. *Journal of Geography, 112*(1), 3–17.

Jo, I., & Bednarz, S. W. (2011). Textbook questions to support spatial thinking: Differences in spatiality by question location. *Journal of Geography, 110*(2), 70–80.

Jo, I., Bednarz, S., & Metoyer, S. (2010). Selecting and designing questions to facilitate spatial thinking. *Geography Teacher, 7*(2), 49–55.

Kastens, K. A., & Liben, L. S. (2007). Eliciting self-explanations improves children's performance on a field-based map skills task. *Cognition and Instruction, 25*(1), 45–74.

Kemp, J., Mellor, A., Kotter, R., & Oosthoek, J. W. (2012). Student-produced podcasts as an assessment tool: An example from geomorphology. *Journal of Geography in Higher Education, 36*(1), 117–130.

Kim, M., & Bednarz, R. (2013). Effects of a GIS course on self-assessment of Spatial Habits of Mind (SHOM). *Journal of Geography, 112*(4), 165–177.

Lanca, M., & Kirby, J. R. (1995). The benefits of verbal and spatial tasks in contour map learning. *Cartographic Perspectives, 21*, 3–15.
Lee, J. (2005). *Effect of GIS learning on spatial ability*. Ph.D. dissertation, Texas A&M University.
Lee, J., & Bednarz, R. (2009). Effect of GIS learning on spatial thinking. *Journal of Geography in Higher Education, 33*(2), 183–198.
Liben, L. S., & Yekel, C. A. (1996). Preschoolers' understanding of plan and oblique maps: The role of geometric and representational correspondence. *Child Development, 67*(6), 2780–2796.
Madsen, L. M., & Rump, C. (2012). Considerations of how to study learning processes when students use GIS as an instrument for developing spatial thinking skills. *Journal of Geography in Higher Education, 36*(1), 97–116.
McGuinness, M. (2009). Putting themselves in the picture: Using reflective diaries in the teaching of feminist geography. *Journal of Geography in Higher Education, 33*(3), 339–349.
Miller, J., Keller, C. P., & Yore, L. D. (2005). Suggested geographic information literacy for K-12. *International Research in Geographical and Environmental Education, 14*(4), 243–250.
Montello, D. R., Lovelace, K. L., Golledge, R. G., & Self, C. M. (1999). Sex-related differences and similarities in geographic and environmental spatial abilities. *Annals of the Association of American Geographers, 89*(3), 515–534.
Olson, J. M. (2006). Map projections and the visual detective: How to tell if a map is equal area, conformal, or neither. *Journal of Geography, 105*(1), 13–32.
Ontario Ministry of Education. (2013). *The Ontario Curriculum, Grades 9 and 10, Canadian and World Studies* (Rev.). Toronto: Queen's Printer.
Schell, E. M., Roth, K. J., & Mohan, A. (Eds.). (2013). *A road map for 21st century geography education: Instructional materials and professional development* (A report from the assessment committee of the road map for 21st century geography education project). Washington, DC: National Geographic Society.
Scholz, M. A., Huynh, N. T., Brysch, C. P., & Scholz, R. W. (2014). An evaluation of university world geography textbook questions for components of spatial thinking. *Journal of Geography, 113*(5), 208–219.
Turner, S., & Leydon, J. (2012). Improving geographic literacy among first-year undergraduate students: Testing the effectiveness of online quizzes. *Journal of Geography, 111*(2), 54–66.
Warkentin, T. (2011). Cultivating urban naturalists: Teaching experiential, place-based learning through nature journalizing in Central Park. *Journal of Geography, 110*(6), 227–238.
Wehry, S., Monroe-Ossi, H., Cobb, S., & Fountain, C. (2012). Concept mapping strategies: Content, tools, and assessment for human geography. *Journal of Geography, 111*(3), 83–92.

Part V
Trends and Recommendations

Chapter 15
Opportunities and Challenges in Using Geospatial Technologies for Education

Joseph J. Kerski

Abstract Opportunities in using geospatial technologies as a meaningful and sustained part of education exist as never before. Most obvious are technological opportunities, centered on the evolution of GST to a web-based platform, including editable online maps that can be customized, analyzed, and shared on any device. Geospatial skills and approaches can be effectively taught in an ever-growing variety of face-to-face and online platforms such as Massively Open Online Courses (MOOCs). But equally important are opportunities in education that focus attention on inquiry, critical thinking, outdoor education, authentic assessment, STEM, technical, green, and other careers, and meaningful teaching with technology. Societal trends offer unprecedented opportunity to use GST in education. These include location analytics that are embedding a positional component in everyday electronic devices, a growing awareness of the geographic significance of key twenty-first century issues, such as biodiversity, urbanization, food, hazards, water, human health, and others, and the increasing role seen for citizen science in solving problems.

Despite these opportunities, technological, pedagogical, and administrative issues still pose challenges. Technological challenges include Internet bandwidth, spatial data availability and permissions, and cost for software as a service. Pedagogical challenges include teaching with rapidly evolving web GST tools and a lack of a "home" for and awareness of geospatial technologies and spatial thinking in the curriculum. Administrative issues such as competition for educational class time and funding pose additional challenges. Nevertheless, geospatial technologies as a "transformational technology" could very well be poised to make a significant positive impact on education and society.

Keywords Opportunities • Challenges • Inquiry

J.J. Kerski (✉)
Esri, Redlands, CA, USA

The University of Denver, Denver, CO, USA
e-mail: jkerski@esri.com

O. Muñiz Solari et al. (eds.), *Geospatial Technologies and Geography Education in a Changing World*, Advances in Geographical and Environmental Sciences,
DOI 10.1007/978-4-431-55519-3_15

15.1 Three Converging Global Trends

Three converging trends offer opportunities to use geospatial technologies (GSTs) and also make a strong case for **why** GST should be used in education (Fig. 15.1). First, there is growing public awareness that the world faces complex challenges that are global in nature but are increasingly affecting our everyday lives. Not a day goes by without the impact of seismic and weather related disasters, such as typhoons, earthquakes, volcanoes, floods, and wildfires, affecting communities or entire countries. Changes in population demographics through birth rates and immigration impact the politics and economics of nations as well as local communities. Sustainable energy supplies are fundamental to life in the twenty-first century and are linked to higher standards of living and educational attainment (Fay et al. 2005; Barnes 1988). Epidemics affect specific segments of society in significant ways. Sustaining agriculture and fisheries, the transportation of food, and water quality and quantity are fundamental to the very existence of humanity. Political instability and violence destroy property and kill and displace whole populations. These challenges have long been some of the fundamental issues that geographers and other earth and environmental scientists have studied. Yet in the past few decades, these challenges have become a part of the everyday public consciousness. These key twenty-first century challenges all occur somewhere, at

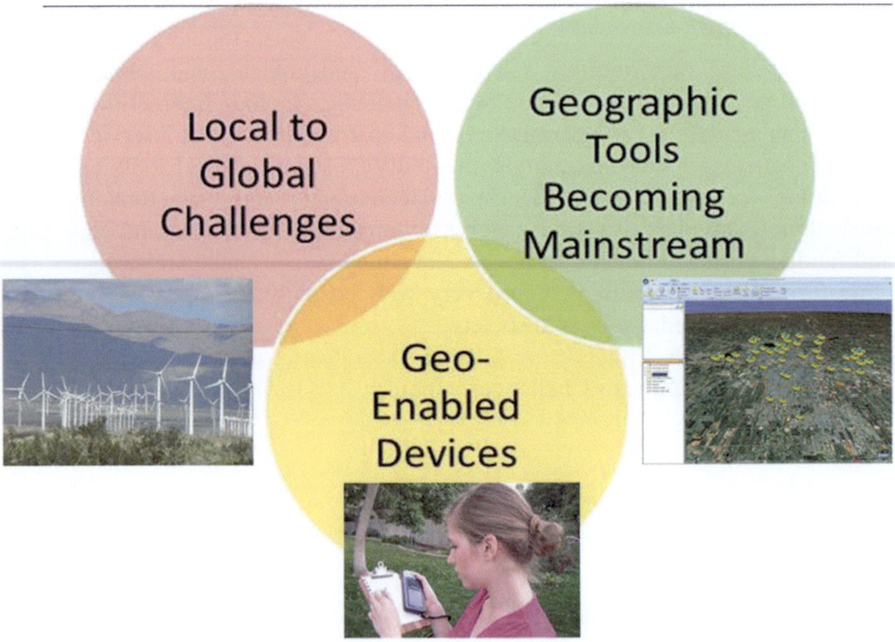

Fig. 15.1 Three trends offering opportunities to use GST that also make a strong case for **why** GST should be used in education (author)

multiple scales, with spatial patterns, with a temporal and a spatial component, and hence can be better understood using the geographic perspective—the "whys of where".

Second, the general public has never before used as many geographic tools as today. Contrast a few hundred cartographers and geographers through the 1980s examining aerial photographs to the number of maps and images that have been viewed by the general public in the time required to read this chapter. Geographic tools, maps, and spatial data can be accessed, streamed, embedded, changed, and re-transmitted on devices from smartphones to tablets to laptops, in the field, vehicles, research labs, classrooms, and everywhere humans are. These digital maps, used in newscasts, web pages, and elsewhere have caused them to be among the most common, valued twenty-first century type of media.

Third, almost everything around us is becoming "geo-enabled." From smartphones to tablet and laptop computers, from webcams recording traffic or bird counts or available parking spaces, from Earth-imaging satellites to on-Earth sensors measuring water quality, all of these sensors and devices transmit a signal, locatable by the coupling of and triangulation of Global Positioning Systems (GPS), smartphone towers, and Wi-Fi transmitters. As geo-enabling extends to thermostats, security systems, and appliances, they become part of "the internet of things" (Wasik 2013) and "smart cities" (Al-Hader and Rodzi 2009) that can all be mapped.

However, the largest part of this sensor network is the general public—over 7 billion strong. As the public uses maps, they voluntarily and involuntarily provide location information through cloud-based smartphone and web-browser applications. Services offer to use this information to make life more efficient, comfortable, interesting. Examples include encouraging others to meet exercise goals through fitness apps, recommending products matching a person's purchasing history, and feeding individuals' speed and location to real-time traffic maps so that other motorists can avoid snarls. This human sensor network together with GST will provide a type of "nervous system" for the planet (Dangermond 2003) analogous to the work of the human nervous system regulating the body's operation.

15.2 Connecting Global Trends to GST in Education

These three trends are opportunities for GSTs in core disciplines and for GST as an essential standalone twenty-first century subject. Furthermore, educational researchers and practitioners recommend that GST be taught often, deeply, and as inquiry (Baker 2005) in problem-based and project-based learning environments (Capraro and Slough 2009), tackling difficult to solve "wicked problems". Inquiry means minimizing fill-in-the-facts worksheets while maximizing hands-on work, discussion, and communication. Inquiry means tackling issues such as the pros and cons of local fracking to issues of rapid growth in the UAE to population decline in Ukraine, and the implications of each. These experiences are tied to field experiences and critical thinking. Others advocate that educators using GST reach beyond

Fig. 15.2 Geoliteracy can be conceptualized as being supported by content knowledge, skills, and the geographic perspective (author)

their own discipline, claiming that the geographic perspective is important from business marketing to biology, economics, history, mathematics, and beyond (Murphy 2007).

Because each complex issue is fundamentally tied to space and place, to grapple with these issues requires a geoliterate population able to assess and use geographic information to make wise decisions (Wertheim et al. 2013). This requires cultivation in each of three legs holding the stool representing geographic literacy: (1) Core content, (2) skills in using geographic tools, and (3) the geographic perspective (Fig. 15.2).

Core content is important though it is frequently maligned, perhaps because it is often equated with memorization of facts for examinations. GST core content extends beyond facts to systems thinking—ecosystems, climate systems, cultural systems, watersheds, ocean systems, land use, political systems, and others. A growing awareness of the centrality of change to twenty-first century life is leading to recommendations that the tools used in education should be those that can model and predict change, using data that changes in real time, and with tools that will themselves update and change. GST meets all of these criteria.

Skills, including the use of GST, are critical to geoliteracy. Many GST tools and skills focus on maps, such as analyzing remotely sensed imagery, using GPS and geolocation, and representing the Earth as map layers. Other critical tools and skills include assessing data quality, graphing, database management, collecting and mapping field data, and communicating.

The geographic perspective is related to “spatial thinking” and has to do with a way of seeing the world. GST enables users to see the world through a series of interwoven, changing spatial relationships, from scale of molecular bonds, to the distribution of macroinvertebrates in water, commuting patterns in a city, seasonal variation in temperate latitudes, to Earth-Sun relationships. The geographic perspective is focused on discovering why processes and phenomena occur where they do, and includes themes of scale, region, diffusion, patterns, and spatial and temporal relationships.

There is also growing awareness that because the public uses geographic tools and data in their everyday lives, they need to understand how to use data and tools wisely. "Wise use" means understanding the capabilities and also the limitations of spatial data, including map projection distortion, spatial and temporal resolution, and attribute completeness. Students graduating with the geographic perspective and skills can inform the public about the appropriate uses of these tools in decision making—in other words, far beyond using maps to navigate to the nearest coffee shop. Because of the geo-enabling of everyday things, those with geographic perspective and skills need to be included in discussions about the wise use of today's deluge of data. These discussions include issues of copyright, privacy, data aggregation, interpretation, dissemination, communication, and implications of that data.

15.3 The Landscape of Geotechnologies in Education: The Plains of Opportunities

15.3.1 The Technological Landscape: Opportunities

Opportunities for the use of GSTs as meaningful components of lifelong education are rapidly expanding. If the use of GST is thought of as a landscape, then the opportunities could be described as plains, easily traversed by educators and students. The most obvious opportunities are those afforded by advances in technology, including GSTs and other tools. I argue that the most significant advancement is the evolution of GIS to a web-based platform. Web-based GIS is a manifestation of the Software-as-a-Service (SaaS) method of delivering software and data through a location hosted online by the providing organization, or, more specifically, the "Cloud." The "Cloud" is a "collection of computers, servers, and databases that are connected together in a way that users can lease access to share their combined power" (Singleton 2011). SaaS is typically accessed by users using a "thin client"–a web browser. Many applications use the SaaS model, including computer games (Minecraft), business (SalesForce), multimedia (Screencast), social media (LinkedIn), and office tools (Dropbox, Microsoft Office 365, Google Drive).

That GST has rapidly embraced the SaaS model has implications for educators, students, and their institutions. Working with GST has always involved working with big data" (Kerski and Clark 2012). For instructors to create GST-based curriculum, data is essential. The cloud offers ease of storage without physical media and larger capacity. To use GST in interdisciplinary problem-solving and learning environments, spatial data needs not only to be stored, but *shared.* Cloud-based GIS data can be streamed or downloaded by multiple individuals and can be simultaneously edited online. Furthermore, not only can spatial data be shared, but so too can the tools, models, workflows, and scripts, fostering a collaborative working environment. This allows the community to spend less time downloading and formatting data, freeing time for spatial analysis, curriculum development, investigating new tools and functionality, and teaching!

One of the chief challenges for educators using technology is reliance on an often overextended Information Technology (IT) staff. This problem is particularly acute with GST because IT staff may hesitate to install something unfamiliar, that updates often, and that which requires great disk space and computing power. With web-based GIS, no software installations are necessary because the software lives online. GST lends itself to teaching with the "Bring Your Own Device" (BYOD) model that has become more commonplace as schools and universities downsize or eliminate traditional computer laboratories. SaaS based data, tools, and capabilities can be accessed and worked with on any device and operating system. Updates are done on the server side, eliminating struggles with keeping one's own "version" up to date.

SaaS online maps are dynamic—customizable, and shareable. They make it easier to map real-time information and incorporate multimedia and fieldwork than desktop-based maps and data. GIS is a system relying upon related skills, tools, and capabilities. GST has grown partly because of advances in computing power, speed, and storage. The internet as a storage and communications medium freed GSTs technically but also socially. Furthermore, as it becomes easier to modify images, edit database fields, embed multimedia into maps, edit maps using HTML and JavaScript, create mashups using Python, host maps using public folders on Dropbox or other platforms, change data formats, upload spreadsheets, geotag videos, map field tracks from GPS receivers and smartphones, instructors' abilities to teach technical skills in tandem with GST is enhanced. GST becomes easier for instructors to justify because it is increasingly seen as *essential to* and *connected with* other twenty-first century tools and skills.

15.3.2 The Instructional Landscape: Opportunities

Policymakers, administrators, and educators have frequently called for increased inquiry-based instruction (Brew 2003; Edelson et al. 1999). Inquiry is at the heart of problem-solving and teaching with GST in the classroom and in informal settings adheres to the tenets of inquiry-driven education (Baker 2005). Emphasis on and funding of STEM education (Science, Mathematics, Engineering, and Mathematics) has lent new attention to interdisciplinary, tech-based methods and technologies, including GST (Nugent et al. 2010).

The advent of educational content standards in many countries during the past few decades and the nationalization of parts of the educational curriculum (Kerr 1990) have challenged and brought opportunities to the use of GST. Emphasis on standardized tests pose a challenge to inquiry driven technology whose benefits are largely in the process, rather than the "single solution" or "one right answer", making the benefits of spatial analysis difficult to assess, fund, and support. However, the focus on problem solving adheres to the tenets of many content standards. Many national content standards include spatial thinking and analysis (Kerski et al. 2013). Some states include GST in their standards and also host courses such as Virginia's Geospatial Semester (Kolvoord 2008).

A focus on twenty-first century skills, including innovation, critical thinking, communication, collaboration, creativity, as well as on career, information, media, and technology (Trilling and Fadel 2009) by key organizations (such as the Partnership for twenty-first century Skills) provide further opportunities. Warnings of the adverse impacts of youth spending too little time outdoors support the use of GST in environmental education (Louv 2006; Kerski 2012; Palmer 2000).

Calls for project-based and problem-based learning also supports the use of GSTs, because these technologies were created to solve problems (Drennon 2005). Authentic assessment, or the practice of evaluating student work based on projects and portfolios rather than standardized tests, also supports the use of educational GST (Solem 2001) because students who solve problems with them are frequently called upon to present their results to their instructor, peers, or a community group.

The engagement that students have with GSTs, cited repeatedly by educators, is a key factor that has moved these technologies forward. Indeed, students using GST exemplify the inquiry process of asking geographic questions, gathering geographic data, assessing geographic information, analyzing geographic information, and acting on the decisions that they make with their new-found knowledge. In a world where citizens bemoan the segmentation of education, GIS brings unifies content, skills, faculty, and students from different disciplines and levels from primary to university (Wright and Goodchild 1997).

The challenge of GST-based curriculum identified by several surveys (Kerski 2003; Baker et al. 2009) has been partly filled by such packages as Mapping Our World, Spatial Mathematics, GeoHistory, In Time and Place, Mapping the Environment, and iGuess. The Body of Knowledge (BoK) built on the earlier Core Curriculum in GIScience by providing a framework upon which to build courses and its revision will continue to do so (DiBiase et al. 2008).

As GST embraced the internet, so too did its professional development, through online courses offered by universities, nonprofit organizations (such as eNet Learning), for-profit professional development companies (such as the Online Learning Consortium), and Massively Open Online Courses (MOOCs). Penn State University's map-focused MOOC attracted 47,000 students in 2013 and in 2014. These represent 'open education' (Sui 2014) opportunities for GST to gain educators, students, and disciplinary areas. In addition, face-to-face opportunities continue to expand, such as the Esri Teachers Teaching Teachers GIS (T3G) annual institutes (Kerski and Demers 2014).

Several significant research advances have aided adoption. These include the Spatial Literacy in Teaching (SPLINT) in the UK (Chalkley 2006), National

Academy of Sciences' *Learning to Think Spatially* report (2006) and the NSF-funded Geography Roadmap project (Wertheim et al. 2013). An online GIS education bibliography topped 2400 entries by 2013 (Baker et al. 2012).

15.3.3 The Societal Landscape: Opportunities

Societal trends have encouraged the use of GSTs in education. These include location analytics that encourage embedding a location in electronic devices, growing awareness of the geographic significance of key twenty-first century issues, growing perception for the importance of fostering good citizenship education, and increasing awareness of roles for citizen science.

Economic challenges and labor shortages in technological fields encourages some countries to embed GST into Career and Technology Education (CTE) programs (Milson et al. 2012). GST was identified in 2004 by the US Department of Labor (Gewin 2004) as one of three major growth fields for the twenty-first century. The NSF funded GeoTech Center effort (Johnson 2010) led community colleges to increase their offerings of GIS certificates and programs (Johnson and Sullivan 2010). The resulting Geospatial Technology Competency Model (GTCM) and guidelines for creating higher education courses aligned with key GST workforce skills (DiBiase et al. 2010; Johnson 2010) provided further opportunity. The GTCM's ethics, organization, and critical thinking support teaching far beyond simply the technical aspects, and its academic competencies of computer science, cartography, geography, earth science, and others reinforce interdisciplinary connections that GIScientists had been building since the early 1990s.

15.4 The Landscape of Geotechnologies in Education: The Rugged Terrain of Challenges

15.4.1 The Technological Landscape: Challenges

Despite opportunities, technological, instructional, and societal issues still pose significant challenges. If opportunities are plains on the GST education landscape, challenges can be thought of as hills when they increase difficulty and crevasses or mountains when they curtail or prevent GSTs from being meaningfully used.

As the internet grows in importance, bandwidth often poses the most significant GST technical challenge. While most school and university campuses around the world offer some sort of internet connectivity, bandwidth required for conducting research and peer collaboration is far less than the bandwidth required for spatial analysis and web mapping. Consequently, not only is cloud-based GST hindered when educational institutions lack bandwidth, but so too is desktop based GST, since so many desktop analysis and basemaps depend upon cloud-based services.

In a survey of educators in 33 countries, the most often mentioned challenge was "access" (Kerski et al. 2012). Not only does access refer to computers with sufficient internal and graphics memory, but to sufficient hard disk space, to software able to handle spatial analysis, and to the school's existing computers. Schools may have well-equipped computer labs, but GST educators often find it difficult to get access to them (Kerski 2003; Baker et al. 2009). Access and support from the school's information technology staff remain challenges in many regions of the world, not confined to the early days of GST (Baker and Kerski 2014) or only to developed countries.

Teaching about GST and teaching with GST has always depended upon suitable spatial data. The open data movement coupled with spatial data infrastructures, data portals, sensors, the ability to quickly map spreadsheets and fieldwork, and georegistering analog maps have vastly increased the amount of spatial data (Kerski and Clark 2012). However, data availability and permissions for its use, particularly at local scales, remains a serious challenge.

Educational institutions pay for running internet services, and if the GST programs are using the bulk of the bandwidth, this may not be viewed favorably by school administrators and accountants. SaaS models often come with costs for accessing geospatial data or services. Educational institutions may be hesitant to adopt this new model, because it is different from the familiar "buy software" model and because subscription costs are partly dependent on difficult-to-estimate usage.

15.4.2 The Instructional Landscape: Challenges

Instructional challenges include the uncertainty of and time required to teach with a rapidly evolving toolset. The paradigm shift in GST today (Reitsma 2013) to web-based GIS creates confusion on what to teach, how to teach, and how best to incorporate. Teaching with GST uses inquiry, typically difficult and time consuming. Its challenges perhaps are best described (and understated!) by Switzerland educators, who said, "Its use in the classroom requires different instruction than is conventionally used (Milson et al. 2012). Thus, professional development is needed, not only in spatial analysis but also in the pedagogical approaches to GIS. It needs to embrace "technological pedagogical content knowledge" (Mishra and Koehler 2006) that captures the interplay among content, pedagogy, and technology.

Lack of research in the implementation and effectiveness of GST in education is a hindrance to its adoption—best practices, how to assess spatial skills, and the difference GST has on student achievement, skills, community connections, and employment. An overreliance on standardized tests that do not assess critical thinking skills constrains any inquiry-based instruction. Thus, the development of national standards, which typically focus on inquiry, has fostered the use of GST, but assessing those standards works *against* the use of GST.

15.4.3 The Societal Landscape: Challenges

Societal issues may cast a greater constraint than hardware and software challenges on GSTs becoming an embedded and required technology throughout education. A lack of awareness of spatial thinking and analysis and their importance in education and society persists. Coupled with the segmentation of modern education into discrete subjects, this translates into a lack of a "home" for GST in the curriculum and, consequently, a lack of funding and support. Geography provides a logical home for GST, and in Turkey, Norway, Taiwan, and the UK, this is where it is most frequently taught (Milson et al. 2012). However, geography is frequently buried in the social studies and, given perennial educational funding crises, faces continual pressure to even maintain the position it has achieved.

That the challenges identified in this chapter have existed for 20 years is frustrating, leading some to ask if the GST education community will be able to overcome what Moore (1999) called "The Chasm" in Rogers' (1995) diffusion of innovations model. If the community has not crossed "the chasm" that would allow GST to attract the "early majority" of educators, what **would** enable the community to cross?

15.5 The Future of GST in Education

Despite the challenges, evidence points to increased future adoption of GST throughout education. The enormous expansion of robust, locatable data will fuel demand for those who can critically work with it. Cloud-based computing will spur those already in GST to become familiar with online platforms such as ArcGIS Online, to use HTML, Python, and JavaScript, and to work with mobile devices, Web APIs, SDKs, and apps. The widening audience for maps as analytical tools (Johnson 2014) means that graduates will need to learn how to communicate with an expanding set of constituents using map-based multimedia such as storymaps.

Increased Internet bandwidth, faster, less expensive computers, and more powerful, easier-to-use GST SaaS and desktop tools will increase adoption. Potential energy shortages may be mitigated by longer-lasting batteries and solar-powered computers. Smartphones reduce the need for a laptop computer. While concerns about the "digital divide" will remain since web-based GST requires a decent Internet connection, initiatives such as "One Laptop Per Child," and bandwidth improvement will ensure that the infrastructure is present in many parts of the world. SaaS GST and voice-enabled software will enable tools in students' native languages, spurring adoption through increased understanding. Increased reliance on home and after-school programs opens new doors to GST (Kerski 2008). Continued curriculum development including the efforts of the Geotech Center (Johnson 2010) will aid adoption. If an international GIS in education research center were created, the resulting research, curriculum, and assessment instruments could connect the community as never before (Baker et al. 2012). But perhaps the largest boost to GST in education will come from citizen science initiatives.

With the convergence of technological, educational, and societal forces, geospatial technologies could make a significant positive impact on education and society throughout the world, becoming a true "transformational technology".

References

Al-Hader, M., & Rodzi, A. (2009). Smart city infrastructure development and monitoring. *Theoretical and Empirical Researches in Urban Management, 2*(9), 87–94.

Baker, T. R. (2005). Internet-based GIS mapping in support of K-12 education. *The Professional Geographer, 57*(1), 44–50.

Baker, T. R., Palmer, A. M., & Kerski, J. J. (2009). A national survey to examine teacher professional development and implementation of desktop GIS. *Journal of Geography, 108*, 174–185.

Baker, T. R., & Kerski, J. J. (2014). Lone trailblazers: GIS In K-12 science education. In J. G. MaKinster, N. M. Trautmann, & G. M. Barnett (Eds.), *Teaching science and investigating environmental issues with geospatial technology: Designing effective professional development for teachers* (pp. 347–372). Dordrecht: Springer.

Baker, T. R., Kerski, J. J., Huynh, N. T., Viehrig, K., & Bednarz, S. W. (2012, Winter). Call for an agenda and center for GIS education research. *Review of International Geographical Education Online, 2*(3).

Barnes, D. F. (1988). *Electric power for rural growth: How electricity affects rural life in developing countries*. Boulder: Westview Press. 236 p.

Brew, A. (2003). Teaching and research: New relationships and their implications for inquiry-based teaching and learning in higher education. *Higher Education Research & Development, 22*(1), 3–18.

Capraro, R. M., & Slough, S. W. (2009). *Project-based learning* (234 p.). Rotterdam. Sense Publishers.

Chalkley, B. (2006). Tertiary teaching excellence: The CETLs scheme in England. *New Zealand Geographer, 62*(2), 155–157.

Dangermond, J. (2003). GIS as a nervous system for the planet. In W. Tang & J. Selwood (Eds.), *Connecting our world: GIS Web services*. Redlands: Esri Press.

DiBiase, D., DeMers, M., Johnson, A., Kemp, K., Taylor Luck, A., Plew, B., & Wentz, E. (2008). Chapter 62: Supporting curriculum development in geographic information science and technology: The GIS&T body of knowledge. In *The manual of GIS* (pp. 1261–1271). Bethesda: American Society for Photogrammetry and Remote Sensing.

DiBiase, D., Corbin, T., Fox, T., Francica, J., Green, K., Jackson, J., Jeffress, G., Jones, B., Jones, B., Mennis, J., Schuckman, K., Smith, C., & Van Sickle, J. (2010, December). The new geospatial technology competency model: Bringing workforce needs into focus. *URISA Journal, 22*(2), 55–72.

Drennon, C. (2005). Teaching geographic information systems in a problem-based learning environment. *Journal of Geography in Higher Education, 29*(3), 385–402.

Edelson, D. C., Gordin, D. N., & Pea, R. D. (1999). Addressing the challenges of inquiry-based learning through technology and curriculum design. *Journal of the Learning Sciences, 8*(3–4), 291–450.

Fay, M., Leipziger, D., Wodon, Q., & Yepes, T. (2005). Achieving child-related millennium development goals: The role of infrastructure. *World Development, 33*(8), 1267–1284.

Gewin, V. (2004, January 22). Mapping opportunities. *Nature, 427*, 376–377.

Johnson, J. (2010, December). What GIS technicians do: A synthesis of DACUM job analyses. *URISA Journal, 22*(2), 41–44.

Johnson, A. B. (2014). Level of knowledge for geospatial occupations and the user community aligned with the competencies covered in higher education courses and programs. GeotechCenter.org.

Johnson, A. B., & Sullivan, D. (2010, December). Geospatial education at U.S. community colleges: Background, challenges, and opportunities. *URISA Journal, 22*(2), 5–14.

Kerr, C. (1990). The internationalization of learning and the nationalization of the purposes of higher education: Two 'Laws of Motion' in conflict? *European Journal of Education, 25*(1), 5–22.

Kerski, J. J. (2003). The implementation and effectiveness of GIS in secondary education. *Journal of Geography, 102*(3), 128–137.

Kerski, J. J. (2008). Towards an international geospatial education community. In *Digital geography: geospatial technologies in the social studies classroom.* (pp. 61–74). Charlotte: Information Age.

Kerski, J. J. (2012). Spatial environmental education: Teaching and learning about the environment with a spatial framework. *Earthzine,* 24 September issue. http://www.earthzine.org/2012/09/24/spatial-environmental-education-teaching-and-learning-about-the-environment-with-a-spatial-framework/. Accessed 1 Oct 2014.

Kerski, J., & Clark, J. (2012). *The GIS guide to public domain data.* Redlands: Esri Press.

Kerski, J. J., & Demers, M. (2014). The Evolution of the T3G Institute and its implications. In *Proceedings, 2014 annual meeting of the Association of American Geographers.*

Kerski, J., Demirci, A., & Milson, A. J. (2013). The global landscape of GIS in secondary education. *Journal of Geography, 112*(6, Fall), 232–247.

Kolvoord, B. (2008). Geospatial technologies: Real projects in real classrooms. *Knowledge Quest 36*(4), 40–45. American Libraries Association.

Louv, R. (2006). *Last child in the woods: Saving our children from nature-deficit disorder.* Chapel Hill: Algonquin Books. 335 p.

Milson, A. J., Demirci, A., & Kerski, J. J. (2012). *International perspectives on teaching and learning with GIS in secondary schools.* Dordrecht: Springer.

Mishra, P., & Koehler, M. (2006). Technological pedagogical content knowledge: A new framework for teacher knowledge. *Teachers College Record, 108*(6), 1017–1054.

Moore, G. A. (1999). *Crossing the chasm: Marketing and selling high-tech products to mainstream customers.* New York: HarperBusiness.

Murphy, A. B. (2007). Geography's place in higher education in the United States. *Journal of Geography in Higher Education, 31*(1), 121–141.

National Academy of Sciences. (2006). *Learning to think spatially: GIS as a support system in the K-12 curriculum.* The National Academies Press. http://www.nap.edu/catalog.php?record_id=11019

Nugent, G., Barker, B., Grandgenett, N., & Adamchuk, V. I. (2010, Summer). Impact of robotics and geospatial technology interventions on youth STEM learning and attitudes. *Journal of Research on Technology in Education, 42*(4), 391–408.

Palmer, J. (2000). *Environmental education in the 21st century.* New York: Routledge. 304 p.

Reitsma, F. (2013). Revisiting the 'Is GIScience a science?' debate (or quite possibly scientific gerrymandering). *International Journal of Geographical Information Science, 27*(2), 211–221.

Rogers, E. M. (1995). *The diffusion of innovations.* New York: Free Press.

Singleton, D. (2011, July 28). What is SaaS? *Software Advice.* http://blog.softwareadvice.com/articles/enterprise/saas-faqs-1072811/

Solem, M. N. (2001). A scoring guide for assessing issues-based geographic inquiry on the world wide web. *Journal of Geography, 100*(2), 87–94.

Sui, D. (2014). Opportunities and impediments for open GIS. *Transactions in GIS, 18*(1), 1–24.

Trilling, B., & Fadel, C. (2009). *21st century skills: Learning for life in our times.* New York: Wiley. 256 p.

Wasik, B. (2013, May 14). Welcome to the programmable world. *Wired.* http://www.wired.com/gadgetlab/2013/05/internet-of-things/

Wertheim, J. A., Edelson, D. C., & the Road Map Project Assessment Committee. (2013). A road map for improving geography assessment. *The Geography Teacher, 10*(1), 15–21.

Wright, D. J., & Goodchild, M. F. (1997). Data from the deep: Implications for the GIS community. *International Journal of Geographical Information Science, 11*(6), 523–528.

Chapter 16
Digital Earth – Digital World: Strategies for Geospatial Technologies in Twenty-First Century Education

Karl Donert

Abstract The chapter introduces the Digital Earth concept and summarizes recent developments in geospatial technology that provide near-ubiquitous access to our Digital World. It identifies the roles played by key stakeholders in the recent, rapid rise and development of the geospatial sector worldwide and considers education from the perspective needs of industry. The chapter then focuses on likely impacts to education and training of the rapid development of open access to geoinformation, geo-Web tools, mobile platforms and Cloud-based technologies. It then considers some issues education and training has been dealing with and maps out some example responses and initiatives. It continues by exploring the importance of sustainable innovation and change in education to meet the needs of industry and society and concludes by suggesting and discussing likely strategies that key stakeholders should implement to enable twenty-first century education to be achieved.

Keywords Digital Earth • Capacity building • Geospatial world

16.1 Introduction

In scientific terms Digital Earth (DE) is a worldwide grouping of scientists, from different disciplines, connected because they are interested in cooperative studies of the planet and its resources, so that they can identify clear, direct solutions for sustainable development. The vision presented by former US Vice President Al Gore (1992, 1998) was that powerful technologies have the potential to create a profound revolution in our understanding of the world. Gore said:

> If we are successful, it will have broad societal and commercial benefits in areas such as education, decision-making for a sustainable future, land-use planning, agricultural, and

K. Donert (✉)
European Association of Geographers (EUROGEO), European Centre of Excellence, digital-earth.eu, Salzburg, Austria
e-mail: kdonert@yahoo.com

O. Muñiz Solari et al. (eds.), *Geospatial Technologies and Geography Education in a Changing World*, Advances in Geographical and Environmental Sciences,
DOI 10.1007/978-4-431-55519-3_16

crisis management. (Gore A. 1998, The Digital Earth: Understanding our planet in the 21st Century, Australian surveyor, 43(2), 89)

In making this statement, Gore considered education and collaborative research to be primary beneficiaries of the technological and societal developments resulting from a Digital Earth initiative.

The DE movement has since made Gore's Digital Earth vision reality. It is an inclusive initiative, which involves many stakeholders including those from education. However, so far, education has not really taken advantage of the high profile, visionary status and prestigious nature of Digital Earth. This chapter examines this by introducing DE as an innovation system and examining the educational opportunities afforded by the Digital Earth initiative. It then introduces examples where and how education has recently responded and advocates for the development of Digital Earth education through continued capacity building in the future.

16.2 Digital Earth: An Innovation System

The Digital Earth initiative demonstrates all the characteristics of an innovation system. According to Koschatzky et al. (2014), an innovation system emphasises the flow of technology and information among people and institutions. Its added value lies in the content, intensity and quality of interactions and networks that result. Johnson and Jacobsson (2000) define seven functions of innovation systems, these are (i) entrepreneurial activity leading to mature technology; (ii) knowledge development through investment, education; (iii) diffusion through networks, events; (iv) guidance from plans and policies; (v) market creation initiatives, (vi) resource mobilization supporting innovation; and (vii) legitimacy through advocacy and leadership. Hekkert et al. (2007) used these functions in mapping innovation systems and concluded that all seven were important variables that influenced their development. Education is therefore a key component of innovation systems, but education systems themselves are marginal in this context as, in general, they tend to have low and slow response rates to change (Burns 2013).

Digital Earth is organized as a loose network of researchers, who advance and utilize geospatial science and geotechnology for the benefit of society. In order to be successful this requires an organizational support infrastructure based on informal networks. The system relies on processes to enable rapid dissemination of knowledge and ongoing formal and informal relationships between individuals and organisations. To achieve this, the International Society for Digital Earth (ISDE) in 2006 was formed to promote academic exchange, DE science and technology innovation. The International Journal of Digital Earth (IJDE) became its main vehicle (http://www.tandfonline.com/toc/tjde20/current#.U-3zM1ZcW0s).

Etzkowitz and Leydesdorff (2000) describe innovation dynamics in terms of communications and negotiations between five groups of key stakeholders:

education, research, policymakers, industry and civil society organisations. They suggest that major efforts are required to develop synergies across these groups to avoid disparate, disconnected developments. Education has become dislocated from Digital Earth multi-stakeholder collaboration. A recent example of this is the Virtual Australia and New Zealand (VANZ) initiative, bringing together research, government, industry, technology and legal stakeholders in Australia and New Zealand (Haines 2014). Its purpose is to negotiate a coherent, consensus-oriented strategy for the development of a 3D computer model of the natural and built environment, inside and outside every structure and utility above and below ground. This is necessary for future decision-making to plan cities and design, test, and market new infrastructure. Education stakeholders were however seemingly not involved even though Digital Earth has undergone a paradigm shift to an open platform that actively supports the possibility of active citizen engagement.

16.3 Digital Earth: Changing Paradigms

The full-scale adoption of Digital Earth has now become envisioned as a tool for policy-making professions, who would become increasingly dependent upon these resources for monitoring and assessment of environmental and social conditions (Craglia et al. 2012). Digital Earth applications are interdisciplinary. Therefore in-depth collaboration between multiple stakeholders and decision-makers from diverse social, economic and political backgrounds is required for knowledge management and knowledge sharing activities (Wright and Wang 2011).

Educators remained a key target audience in the Digital Earth use cases proposed by Grossener et al. (2008). Educational potential was demonstrated through metaphors like Geo-library and Virtual Atlas as context scenarios. However, more recently, a rift between Digital Earth science and Digital Earth education has developed largely as a result of the chasm between the operational environments of educational policy makers, DE practitioners, the geospatial industry and researchers. Despite many national initiatives (Milson et al. 2012), education frameworks around the world have not been able to keep pace with rapid DE advances. Nevertheless, as Gould et al. (2008) recommended, education should not be forgotten, as the Digital Earth innovation system requires investment in capacity building instructional programmes.

Education has become less important in recent Digital Earth policy priorities. Indeed, in the most recent vision of Digital Earth, the role of education is no longer mentioned (Goodchild et al. 2012). Now Digital Earth is envisaged as a common platform to support national and international cooperation for global, sustainable development and as a newly developing point of economic growth and social wellbeing (Ehlers et al. 2014). Citizens are the new target group, with rights of access to information made possible by combining Web 2.0, Digital Earth and mobile technologies for the democratisation of data for all. The widespread adoption of social media also turns citizens into major providers of information and affords opportunities for the gathering and utilisation of big data sets. Goodchild (2007) termed this "Citizens as Sensors", where volunteered geographic information transforms Digital Earth and the ways social mobilization occurs. Citizen Science, as a form of social computing, plays an increasing role in the context of people used as sensors. Citizens participate in or initiate the collection and sharing of measurements of their everyday environments. So far, these possibilities have hardly permeated into even the most forward-looking educational situations.

16.4 Digital Earth Opportunities for Education

The components of Digital Earth computing platforms have become available to educators through the many advances in aspects of functionality, user interface and system architecture (Henfridsson et al. 2014). There have been major efforts to build the technical capability to serve geographic media (written also as geo-media) at a massive scale, facilitated by developments in Web 2.0, Cloud computing and communications technologies. The potential of these technologies is so powerful that its advances have led to an explosion in the number of freely available geospatial tools, apps, interactive maps and mashups. These make massive amounts of geo-media freely available for learners, teachers and educators to access, use, share and communicate, as advocated by communication theory (Thompson 2013) and contemporary pedagogy for a digital age (Beetham and Sharpe 2013).

Geo-media is a relatively new concept linking geoinformation, geotechnology and multimedia representation (Gryl et al. 2010). It is increasingly being used as a term in school education publications (de Miguel 2012; Gryl and Jekel 2012; Roosaare and Liiber 2013). It is described as (multi-)media derived from spatially enabled mobile devices, online mapping tools and even as volunteered geographic information. It relates to any form of media that incorporates or portrays geographical (location-based) information. It links people to real-world issues by socially connecting them through geographic location and the Geospatial Web (Elwood 2010). However, with such a variety of tools continually emerging and the ubiquitous nature of geo-media, education systems have been finding it too challenging to keep pace with the changing nature of twenty-first century learning and the many opportunities on offer (De Miguel and Donert 2014).

In education, Digital Earth provides countless opportunities for innovative learning and teaching developments in most subject areas. For instance, DE can help provide a vision of the future of our planet, virtually representing it in terms of its systems and forms (Hicks 2013). It offers the potential of individual and collective engagement through challenging, deeper forms of learning by encouraging analysis, critical thinking and scenario building. So, learning environments using Digital Earth technologies and geo-media help establish authentic, profound learning experiences, while raising levels of engagement, participation and understanding. This potential warrants the development in education of new capabilities if people are to successfully participate as responsible "spatial citizens" (de Miguel 2012), as advocated by the Spatial Citizenship (SPACIT) Project (http://www.spatialcitizenship.org).

SPACIT is currently exploring innovative approaches to the inclusion of geo-media in schools, as part of the process of engaging young people in their own environments (De Luca et al. 2014). It has produced a competence model (Gryl et al. 2013) and curriculum (Schultz et al. 2014) with the goal to develop online materials for teacher education and training. Key to spatial citizenship is positive uses of geo-media in education and the integration of geospatial tools in everyday life. This is possible through the almost ubiquitous availability of smartphones, tablets, wifi and the Cloud. Spatial citizenship implies bridging the gap between citizens and democratic processes through geo-media, encouraging pro-active participation and the appropriation of space (Gryl et al. 2010). Spatial citizenship demonstrates the capacity to create powerful learning opportunities that empower young people and individualise learning (Hogrebe and Tate 2012).

16.5 Capacity Building the Geospatial Industry

Future Digital Earth innovation will most probably come from the private sector (Goodchild et al. 2012). There is general recognition by these stakeholders that DE evolution needs to be better integrated into education in order to meet both the present and the future needs of the geospatial industry (Tsai and Chen 2014). Currently this is not taking place and the level of necessary skills and training opportunities does not meet current demand (GeoSkills Plus 2014). Significant capacity building in learning and teaching across all levels of education is therefore essential for this to happen. O'Brien et al. (2013) comment that this will require more than slight adjustments in current educational systems, funding strategies, and collaboration. They suggest trans-disciplinary approaches to learning, teaching and education in general are required, as well as the development of necessary capabilities.

Giuliani et al. (2013) identify three components in capacity building: (i) providing a necessary infrastructure, (ii) improving the understanding of the value of geospatial data to support decision-making and (iii) having suitable levels and quality of education and training to meet geo-labour market demands. These actions, while also implementing necessary changes in education to effectively respond to actual environmental and social challenges, will also stimulate further growth through the endorsement of Digital Earth technologies. Industry believes education needs to adopt more innovative approaches to embrace the rapidly changing scientific, geospatial education landscape.

In the meantime, Kumar (2013) illustrates how demand for geospatial products and related services continues to increase, with the need for more highly skilled people not only in traditional disciplines like surveying and cartography but also in growth areas like big data, location based services and remote sensing. There is now a real risk that major skills shortages in these areas will limit geospatial industry growth and its ability to meet changing and growing demands for services and products in areas such as land markets, environmental management, utilities, national security and transportation (GeoSkills Plus 2014). Shortages of skilled and creative people also limit the ability to develop innovative market-oriented spatial services in a world where applications of spatial information are becoming pervasive.

In response to these needs, some large-scale geospatial capacity building education actions have been initiated in different parts of the world. In India, realizing the significance of geospatial training, a national Task Force was established to advise on a national geospatial strategy to implement the vision for rapid, sustainable and more inclusive growth. They made four specific recommendations: (i) a school geospatial education project, (ii) geospatial university networking (iii) geospatial training and outreach; and (iv) a National Institute of Geospatial Knowledge, Engineering and Management (MHRD 2013).

In the USA, the social and financial implications of geospatial technology growth has prompted the government to identify GIS as one of the three most important emerging and evolving fields, with an estimated 950,000+ additional jobs in geospatial technology expected between 2005 and 2015 (U.S. Department of Labor 2011). The National Science Foundation awarded in excess of $20 million for research in geospatial technologies (GSTs) in education. Recent activities (May 2014) confirmed President Obama's personal involvement at the launch of a $1 billion "geospatial education for all" initiative to help strengthen STEM education for school students (GeoSpatial Solutions 2014).

In Europe, GSTs are expected to experience an 11 % compound annual growth rate between 2014 and 2016 (Daratech 2012). The European Commission has paid close attention to evolving technological shifts, as the foundation of the social, cultural, economic and political future of Europe. No unified policy for geospatial capacity building has so far been suggested. Instead the wider uses of ICT and data are stressed, without explicit reference to geospatial technology or geoinformation. As a result, the geospatial workforce remains undefined as an employment category in most European countries, the evidence necessary to support capacity building is not being gathered and educational projects funded under relatively short-term timeframes dominate.

One significant European development, in response to geospatial science advances, was the foundation of an Austrian Centre for geospatial media education (digital:earth:at) created in Salzburg in 2009. This linked a number of Austrian organizations working with schools and teachers and connected them with scientists and researchers (Jekel et al. 2012). The goal was to share resources, tools and innovative ideas and increase the use of geospatial science with pupils and teachers. Its implementation led to the development of a European networking initiative, called "digital-earth.eu – geo-media in schools for a better world". This project connected education stakeholders and promoted geospatial media tools, data and learning opportunities. A partnership of 87 organizations from more than

20 countries was formed and funding obtained for them to work together for 3 years (2010–2013) under the EC Lifelong Learning Programme (De Miguel and Donert 2014).

In order to build capacity as part of this project, a European Centre was formed to establish a European infrastructure of expert Digital Earth Centres (http://www.digital-earth-edu.net) with the potential to form multipliers in different countries by working with many teachers and trainers in their own contexts. This encouraged them to scale-up Digital Earth education and training, influence national and European policy and communicate and advocate future visions as part of an informed Community of Practice. Recognition offers those involved visibility and encourages further innovation and entrepreneurial activity, through spinoff curriculum, teacher training and careers projects like Spatial Citizenship (http://spatialcitizenship.eu), GeoSkills Plus (http://geoskillsplus.eu) and I-Use for statistics in schools (http://www.i-use.eu).

Without sufficient capacity building, education for Digital Earth will not become prominent. At the 2014 Geospatial World Forum workshop on GeoCapacity Building, participants were asked by the author to consider and share their opinions on what should be done to capacity build geospatial industry. The main strategies identified were to: (a) network industry with education, by establishing a leaders forum; (b) better promote geospatial professions through a careers portal; (c) cluster and support good pilot projects to make them sustainable with the help of industry; (d) build research capacity to better understand labour market supply and demand; and (e) create qualifications benchmarks, enabling more workforce mobility (http://tinyurl.com/GWF-geocap2014). All of these different geo-capacity building strategies are much needed, but remain challenging for education and industry to deliver.

16.6 Conclusions

Digital Earth has become reality and its outcomes should play an increasingly important role in education as the science is used to address the social, economic, cultural, scientific, and technological challenges affecting the way we understand and interact with the earth. Digital Earth presents an excellent opportunity to make meaningful connections between spatial phenomena and geographical processes, the tools of visualization and representation and their applications in everyday life (Kavouras et al. 2014).

Digital Earth demonstrates many characteristics of an emerging innovation system (Bartels et al. 2012). Developments have created a profound revolution in science and technology, information access and spatial thinking. Advances are accelerating rapidly and the world of technology and industry seems always to be more than one step ahead of educational institutions. The rising tide of geospatial technical change cannot be controlled, but related educational initiatives generally lack the ability to respond due to insufficient funding. The previously highlighted initiatives show that Digital Earth education has started responding to the challenge of capacity building, but this needs much more support from industry to further innovative developments (NGAC 2012). The scaling up and sustainability of pioneering teacher education and training initiatives like the SPACIT Project must be further encouraged.

In future, education developments related to Digital Earth should increase their visibility through involvement with other major international activities, such as the UN International Year of Global Understanding (IYGU). This international year, proposed by the International Geographic Union, addresses the ways in which we inhabit our increasingly globalized world (http://www.global-understanding.info/). Education for Digital Earth fits effectively within the IYGU remit and outcomes as it contributes to the ICSU-ISSC Future Earth initiative and supports the implementation of the Rio+20 Declaration.

Volunteer initiatives centred on open-source tools and geospatial data regimes, such as the Open Geospatial Consortium (www.opengeospatial.org), have been very successful. Kerski (2008) called for the establishment of a Digital Earth community of educators. As GSTs are now technologically mature and their use is becoming more widespread, perhaps it is now time to initiate this. Widespread collaboration under a Digital Earth banner, among stakeholders interested and active in education, could be successful in developing 'Digital Earth education for all'.

The main question that remains is who will take responsibility for geo-capacity building in education for Digital Earth? While a plethora of global initiatives already deal with different aspects of information management and new technologies, stakeholder initiatives like EyeonEarth and 'Geo for All' promote open access to geospatial information, no such high-level body addresses capacity building for Digital Earth education. This remains a clear gap that needs to be filled by strong leadership in geospatial education.

References

Bartels, F. L., Voss, H., Lederer, S., & Bachtrog, C. (2012). Determinants of national innovation systems: Policy implications for developing countries. *Innovation: Management, Policy & Practice, 14*(1), 2–18.

Beetham, H., & Sharpe, R. (Eds.). (2013). *Rethinking pedagogy for a digital age: Designing for 21st century learning*. London: Routledge.

Burns, M. (2013). Success, failure or no significant difference: Charting a course for successful educational technology integration. *International Journal of Emerging Technologies in Learning, 8*(1), 38–45.

Craglia, M., de Bie, K., Jackson, D., Pesaresi, M., Remetey-Fülöpp, G., Wang, C., et al. (2012). Digital earth 2020: Towards the vision for the next decade. *International Journal of Digital Earth, 5*(1), 4–21.

Daratech. (2012). GIS/geospatial industry worldwide growth slows to 1 % in 2009. *Directions Magazine*. http://www.directionsmag.com/pressreleases/gisgeospatial-industry-worldwide-growth-in-2012/. Accessed 5 May 2014.

De Luca, N., Jekel, T., & Ferber, N. (2014). Spatial citizenship education – Empowering citizens with geo-media. *INTED2014 Proceedings* (pp. 1618–1627).

De Miguel, R. (2012). Geomedia for education in sustainable development in Spain, an experience in the framework of the aims of digital-earth.eu. *European Journal of Geography, 3*(3), 44–56.

De Miguel, R., & Donert, K. (Eds.). (2014). *Innovative learning geography in Europe: New challenges for the 21st century*. Newcastle: Cambridge Scholars Publishing.

Ehlers, M., Woodgate, P., Annoni, A., & Schade, S. (2014). Digital earth reloaded–Beyond the next generation. In *IOP Conference Series: Earth and Environmental Science*, 18(1), IOP Publishing

Elwood, S. (2010). Geographic information science: Emerging research on the societal implications of the geospatial web. *Progress in Human Geography, 34*(3), 349–357.

Etzkowitz, H., & Leydesdorff, L. (2000). The dynamics of innovation: from national systems and "Mode 2" to a Triple Helix of university–industry–government relations. *Research Policy, 29*(2), 109–123.

GeoSkills Plus. (2014). *Project description*. http://www.geoskillsplus.eu/project-description/. Accessed 29 May 2014.

GeoSpatial Solutions. (2014). *Jack Dangermond of Esri pledges $1B of mapping software to schools*. http://geospatial-solutions.com/jack-dangermond-of-esri-pledges-1b-of-mapping-software-to-schools/. Accessed 6 June 2014.

Giuliani, G., Ray, N., & Lehmann, A. (2013). Building regional capacities for GEOSS and INSPIRE: A journey in the Black Sea catchment. *International Journal of Advanced Computer Science and Applications, 3*(3), 19–27.

Goodchild, M. (2007). Citizens as sensors: The world of volunteered geography. *GeoJournal, 69*(4), 211–221.

Goodchild, M. F., Guo, H., Annoni, A., Bian, L., de Bie, K., Campbell, F., & Craglia, et al. (2012). Next-generation digital earth. *Proceedings of the National Academy of Sciences, 109*(28), 11088–11094. http://www.pnas.org/content/109/28/11088.full. Accessed 2 June 2014.

Gore, A. (1992). *Earth in the balance*. New York/Houghton: Mifflin Company.

Gore, A. (1998). The digital earth: Understanding our planet in the 21st century. *Australian surveyor, 43*(2), 89–91.

Gould, M., Craglia, M., Goodchild, M. F., Annoni, A., Camara, G., Kuhn, et al. (2008). Next-generation digital earth: A position paper from the vespucci initiative for the advancement of geographic information science. *International Journal of Spatial Data Infrastructures Research, 3*, 146–167.

Grossner, K. E., Goodchild, M. F., & Clarke, K. C. (2008). Defining a digital earth system. *Transactions in GIS, 12*(1), 145–160.

Gryl, I., & Jekel, T. (2012). Re-centring geoinformation in secondary education: toward a spatial citizenship approach. *Cartographica: The International Journal for Geographic Information and Geovisualization, 47*(1), 18–28.

Gryl, I., Jekel, T., & Donert, K. (2010). GI and spatial citizenship. *Learning with GeoInformation, V*, 2–11.

Gryl, I., Schulze, U., & Kanwischer, D. (2013). Spatial citizenship: The concept of competence. In R. Vogler, A. Car, J. Strobl, & G. Griesebner (Eds.), *Creating the GI Society* (pp. 282–293). Berlin: Wichmann.

Haines, M. (2014). Virtual Australia and New Zealand (VANZ): Creating a piece of digital earth. In *IOP Conference Series: Earth and Environmental Science*, 18(1), 012006, IOP Publishing. http://iopscience.iop.org/1755-1315/18/1/012006/pdf/1755-1315_18_1_012006.pdf. Accessed 6 June 2014.

Hekkert, M. P., Suurs, R. A. A., Negro, S. O., Kuhlmann, S., & Smits, R. E. H. M. (2007). Functions of innovation systems: A new approach for analyzing technological change. *Technological Forecasting & Social Change, 74*, 413–432.

Henfridsson, O., Mathiassen, L., & Svahn, F. (2014). Managing technological change in the digital age: the role of architectural frames. *Journal of Information Technology, 29*, 27–43.

Hicks, D. (2013). Developing a futures perspective in the classroom. In S. Ward (Ed.), *A student's guide to education studies* (pp. 133–145). London: Routledge.
Hogrebe, M. C., & Tate, W. F. (2012, March). Geospatial perspective: Toward a visual political literacy project in education, health, and human services, *Review of Research in Education, 36* (1): 67–94.
Jekel, T., Koller, A., & Strobl, J. (2012). Austria: Links between research institutions and secondary schools for geoinformation research and practice. In A. J. Milson, A. Demirci, & J. J. Kerski (Eds.), *International perspectives on teaching and learning with GIS in secondary schools* (pp. 27–36). Dordrecht: Springer.
Johnson, A., & Jacobsson, S. (2000). Inducement and blocking mechanisms in the development of a new industry: The case of renewable energy technology in Sweden. In R. Coombs (Ed.), *Technology and the market: Demand, users and innovation* (pp. 89–111). Cheltenham: Edward Elgar Publishing.
Kavouras, M., Sotiriou, S., Baglatzi, A., Darra, N., Kokla, M., & Lazoudis, A., et al. (2014). Semantic pathways for geospatial thinking. In Stracke, C. M., Ehlers, U. D., Creelman, A. & Shamarina-Heidenreich, T. (Eds.), *Changing the trajectory: Quality for opening up education, Efquel.* http://cdn.efquel.org/wp-content/blogs.dir/2/files/2014/05/EIFLINQ_2014_Proceedings.pdf Accessed 21 May 2014.
Kerski, J. J. (2008). The role of GIS in Digital Earth education. *International Journal of Digital Earth, 1*(4), 326–346.
Koschatzky, K., Schnabl, E., Zenker, A., Stahlecker, & T., Kroll, H. (2014). *The role of associations in regional innovation systems* (No. R4/2014). Working Papers Firms and Region. http://www.econstor.eu/bitstream/10419/95962/1/782711464.pdf. Accessed 8 June 2014.
Kumar, P. (2013). *GIS market growth during 2012–2016*, GeoAwesomeness. http://geoawesomeness.com/gis-market-growth-2012-2016/. Accessed 21 May 2014.
MHRD. (2013). National task force report on geospatial education, Ministry of Human Resource Development: Government of India, https://nsdiindia.gov.in/nsdi/nsdiportal/meetings/MHRD-National%20Task%20Force%20on%20Geospatial%20Education-Vol%20I%20&%20II%20Combined.pdf. Accessed 10 June 2014.
Milson, A. J., Demirci, A., & Kerski, J. J. (Eds.). (2012). *International perspectives on teaching and learning with GIS in secondary schools*. New York: Springer.
NGAC. (2012). Geospatial workforce development, National Geospatial Advisory Committee. http://www.fgdc.gov/ngac/ngac-geospatial-workforce-development-paper-final.pdf. Accessed 5 June 2014.
O'Brien, K., Reams, J., Caspari, A., Dugmore, A., Faghihimani, M., Fazey, I., et al. (2013). You say you want a revolution? Transforming education and capacity building in response to global change. *Environmental Science & Policy, 28*, 48–59.
Roosaare, J., & Liiber, Ü. (2013). GIS in school education in Estonia–looking for an holistic approach. *J-Reading-Journal of Research and Didactics in Geography, 1*(1), 47–56.
Schulze, U., Gryl, I., & Kanwischer, D. (2014). Spatial citizenship: Creating a curriculum for teacher education. In R. Vogler, A. Car, J. Strobl, & G. Griesebner (Eds.), *Geospatial innovation for society* (pp. 230–241). Berlin: Wichmann.
Thompson, J. B. (2013). *Media and modernity: A social theory of the media.* Cambridge: Polity Press.
Tsai, F., & Chen, L. C. (2014). Challenges of remote sensing and spatial information education and technology transfer in a fast developing industry. *ISPRS-International Archives of the Photogrammetry, Remote Sensing and Spatial Information Sciences, 1*, 103–106.
U.S. Department of Labor. (2011). *Employment and training administration, career one stop, competency model clearinghouse, geospatial technology.* http://www.careeronestop.org/CompetencyModel/pyramid.aspx?GEO=Y&ES=Y&EST=geospatial. Accessed 6 June 2013.
Wright, D. J., & Wang, S. (2011). The emergence of spatial cyberinfrastructure. *Proceedings of the National Academy of Sciences, 108*(14), 5488–5491.

Chapter 17
An Agenda of GST in Geography Education for the Future

Injeong Jo and Osvaldo Muñiz Solari

Abstract Recent advances in geospatial technologies (GST), wide availability of geospatial data, and a growing interest in the use of GST in education call for a comprehensive model for teaching and learning with GST. We propose a curricular model that can serve as a guide for integral geography education with GST, where spatial abilities and geospatial integrative skills constitute the two key pillars. Learning opportunities for students to acquire spatial concepts knowledge and to practice higher-level cognitive processes with GST must be critical components of this new approach. Instructional and assessment practices using GST should be designed in a way that can facilitate spatial thinking and geographic inquiry. Focusing on students' understanding of concepts, ability to utilize a variety of tools of representations, and practice of problem solving through spatial analysis and reasoning in an authentic context can be an effective strategy. A challenge is preparing teachers who are capable of designing and implementing GST-supported inquiry-based instruction and authentic assessment. More research that focus on knowledge, skills, and dispositions related to teaching with GST should shed light on this matter.

Keywords Geospatial education • Geospatial curriculum • Geospatial instruction

17.1 Introduction

From Web 1.0, a read-only information flow, to Web 2.0 (O'Reilly 2005), geospatial sciences and particularly geography have begun to take advantage of geospatial technologies (GSTs) along the read-write Web. The profusion of user-generated content and dynamic information flow allow innovative ways of geospatial practice (GP) with uninterrupted relational problems. Influenced by these new trends of technological prospect and fresh geospatial skills, geography

I. Jo (✉) • O. Muñiz Solari
Department of Geography, Texas State University, San Marcos, TX 78666, USA
e-mail: ijo@txstate.edu; o.muniz@txstate.edu

O. Muñiz Solari et al. (eds.), *Geospatial Technologies and Geography Education in a Changing World*, Advances in Geographical and Environmental Sciences,
DOI 10.1007/978-4-431-55519-3_17

educators' wide research future is required to investigate the effectiveness of new models for geospatial education.

Reorganization of geographic education curriculum and pedagogy is a great challenge in the continuous and permanent process to prepare spatially literate participants. They are demanding better and deeper knowledge and skills; an evident outcome of the information technology based upon the Web. Guiders' abilities to work with GSTs and expedite GPs have to be initiated by willingness and inclination to incorporate a new approach. Technological Pedagogical Content Knowledge (TPACK) (Mishra and Koehler 2009) might help in this direction. More collaborative procedures for assimilation of new integrative knowledge and geospatial skills must be ruled across the curriculum.

Instruction should give increasing space to facilitation in the dynamic pedagogical process of learning about the Earth with GST integrative skills. Geography educators are responsible for guiding in the process of understanding and mastering endless geographic spatial relationships which are boosted by growing complex data. With lightweight programming models served through Cloud and multiple GST devices, geography education needs to enter a new arena characterized by integrative spatial thinking and integrative inquiry-based learning.

17.2 Novel Model to Facilitate Geospatial Capacity Building

The complexity of our tasks to understand and master the earth dynamics impels us to formulate a basic agenda of geospatial science for geography education. It focuses on the preparation of new generations of GST practitioners who are exposed to GP through either formal education or non-formal and informal settings. This agenda should assist educators in the process of guiding the good use of GST and the exercise of GP in order to produce awareness of local, regional as well as global issues which concern mankind as a whole.

GST must be considered as mediums for teaching as Mishra and Koehler (2006) see the incorporation of new technologies in education. Moreover, GST constitutes an intrinsic part of the professional preparation in geographical sciences and related fields today. The worldwide scientific community is becoming well aware of this fact (Tang et al. 2008; Aina 2012; Tejpal et al. 2012; Vyas and Koening 2014). Therefore, school teachers, college instructors, and university professors are increasingly sensitive about new teaching and learning procedures to study geographic phenomena and resolve geographic problems with GST (De Vecchis et al. 2011; Tang et al. 2008; Lisec and Ruiz Fernandez 2008). It is not a technological option or mode that can be replaced by other mediums to study the earth phenomena. GST is an integrated body of technologies (Gao 2002; Yang et al. 2010) created and developed to measure complex earth's processes by using quantitative and qualitative procedures (Sui and DeLyser 2012).

The increasing use of GST among practitioners implies a pedagogical transformation in formal and non-formal education, which has almost similar characteristics in informal learning. Attributes of this pedagogical transformation have been widely studied in different countries. Active learning (Scheyvens et al. 2008; Tuna 2012), collaborative structure (Hertzog and Lieble 1996; Muñiz-Solari and Coats 2009), problem solving (Schneider-Sliwa and Leser 2003; Pawson et al. 2006), project design (Livingstone and Lynch 2000; De Vecchis et al. 2011; Demirci et al. 2011), and community service (Hagevik 2011; Hauselt and Helzer 2012) are among the most important attributes.

17.2.1 First Component of the Model: Spatial Abilities

Several indications presented in different chapters in this book announce the importance of spatial thinking as one of the fundamental abilities to be performed while using GST and being engaged in GP. Hence, spatial abilities are the first components that need to be reviewed as part of the agenda of geospatial science in geography education for the future. Spatial literacy and spatial thinking form two critical pillars of the geospatial capacity building. Spatial literacy, defined as understanding of spatial concepts, is fundamental to develop good spatial thinking and subsequent geospatial thinking.

The open access within the geo-enabling world has paved the way to the implementation and wide availability of GST to anyone who has acquired certain level of training with these technologies. Yet an increasing number of practitioners of GST are not equivalent to a geospatially-literate population. Even when the understanding of fundamental spatial concepts varies according to different cultural settings and their conditions, there are clear evidences that geographic illiteracy is obvious among young population engaged in formal education around the world (Misheck et al. 2013; Schoeller et al. 2011). It is either a partial or total lack of systematic and progressive knowledge on spatial concepts and vocabulary that affect the literate level of practitioners.

The organization of basic concepts (i.e., location, direction, distance) and advanced concepts (i.e., nodes, network, diffusion) should form the solid structure of spatial literacy. This structure has to be built by young generations to initiate their knowledge on GST and make progress toward the practice of integrated GST. Golledge et al. (2008) introduces a 'spatial task ontology' and concept base that facilitate a foundation for geospatial thinking. Given the experiments developed by the researchers we propose to follow a similar sequence of concepts as a starting point. The application of the five-level ontology of primitive, simple, difficult, complicated and complex terminology should be the initial framework for developing school geography curricula. Geospatial tasks, embedded concepts and more frequent geospatial applications should be exercised for enabling understanding geospatial relationships.

We have to recognize that among the most common GSTs only Geographic Information System (GIS) has been more developed in the schools (Kerski et al. 2012). GIS is a task-based system as they are all the rest of GST used by a great variety of professional practitioners. The need for a task-based ontology to assist school students and teachers in their ability to use concepts is also evident. GIS instruction and concept difficulty are two combined issues that demand a great deal of attention as some researchers have discovered (Huynh and Dean 2011). In relation to this ultimate objective it is important to recognize the level of difficulty practitioners (young and adults) in both formal and informal learning may encounter. Difficulty means not only to understand the concepts in their most simple definition but also find the ability to use and combine numerous concepts to resolve some problem. Consequently, it is necessary to develop a geospatial concept lexicon and task ontology that provides a precise plan for teaching and learning processes on GST.

The spatial literacy will not be complete if we do not tackle the highest level of conceptualization in order to use GSTs with efficiency. Threshold concepts allow practitioners to transition from advanced conceptualization to integrative meaning or mental models (Meyer and Land 2005; Timmermans 2010). The integrative power of threshold concepts enables understanding linkages among different basic and advanced concepts (Srivastava 2012). Threshold concepts identification and its use in curriculum design promote research collaboration among students and instructors (Cousin 2010). Furthermore, as Cousin points out, it resolves the dilemma of student-center versus teacher-center approach. Neither teacher nor student is the center but both working in collaboration. In any case, the new curriculum for geospatial literacy in formal education requires a progressive development on conceptualization that has to be consistent with the level of complexity GST and GP need. This complexity is intimately related to spatial thinking.

Conceptualizations of spatial thinking have been presented by Metoyer, Bednarz and Bednarz in Chap. 3, among which the interactionist approach accepts differentiation in the process of initial steps people use to develop spatial thinking. The influence of culture and cultural tools is present (Newcombe 2000) and some researchers suggest cultural dissimilarities (Oyserman and Lee 2008), while others indicate common cognitive norm by people of different cultures. For instance, topological thought is important in GP, and Knauff (2013) points out that there are cross-cultural similarities regarding this spatial reasoning. However, spatial skills are enhanced by training which is initiated with spatial concepts.

We propose an agenda to activate a curriculum that develops a sequence of intensive applied activities with GST based on three orders of spatial thinking. They are based on the National Research Council's original proposition as learning process to think spatially (National Research Council [NRC] 2006): (1) Practice with inner spaces which are also recognized as personal life spaces (3rd Order of spatial thinking), (2) Practice with external or physical spaces (2nd Order of spatial thinking), and (3) Practice with abstract spaces referred to as intellectual constructs (1st Order of spatial thinking). Therefore, practices with GST should be related to spatial thinking *in* space, *about* space, and *with* space, respectively. Tools for spatial thinking and reasoning must be organized by an increasing engagement with GST; both as instruments and set of techniques.

It has been proposed that the three most important components of spatial thinking are concepts of space, tools of representation, and processes of reasoning (NRC 2006). The first component has already been analyzed as part of the agenda for GST. The second component is represented by several forms of communication such as maps, models, diagrams, graphs, sketches, and charts. Tversky (2005) indicates the importance of these tools for spatial thinking and reasoning. Yet the accuracy obtained with these tools is not always perfect; issue that leads us to the third component.

Processes of reasoning are, in fact, cognitive processes which are vital components of spatial thinking. First, it is important to determine the best taxonomy of cognitive processes in formal education to easy the process of spatial thinking with GST. We propose the use of Morseley's cognitive skills (Morseley et al. 2005) organized in the cognitive domains (i.e., information gathering, building understanding, productive thinking) as the most appropriate to work with GST and develop GP. They should be implemented as part of the agenda to develop spatial abilities. As for the spatial thinking it is important to understand the new structure of the spatial world to use broad categories of spatial skills. A recent typology presented by Newcombe and Shipley (2014) gives us an organizing schema for thinking the spatial skills that might be important for GST practitioners in K-16 education working in a geo-enabling world. It is derived from cognitive, linguistic, neural, computational, and STEM learning perspectives. Four cells of this typology (i.e., intrinsic-static skills, intrinsic-dynamic skills, extrinsic-static skills, extrinsic-dynamic skills) provide a wide variety of spatial skills. Large scale spatial skills, navigation skills, mental rotation skills, folding skills, cross-sectioning skills, etc. represent typical examples of skills that are required for GST training characterized by relational reasoning. Ultimately it is a process of migration from spatial thinking to geospatial thinking that is immersed in the first component of the model presented by Muñiz-Solari (Fig. 17.1-A). The second component (B); geospatial integrative skills, shows the main elements that are part of our next topic.

17.2.2 Second Component of the Model: Geospatial Integrative Skills

The complexity of our geospatial problems in the world is leading us to prepare students at their early age in formal education to deal with an integrative perspective. In terms of infrastructure geography teachers and students need integrative skills to deal with data storage and retrieval as well as cyberinfrastructure and tools (NRC 2010).

Data handling – data sharing and technology integration are the main factors of the second component recognized as geospatial integrative skills. Dealing with

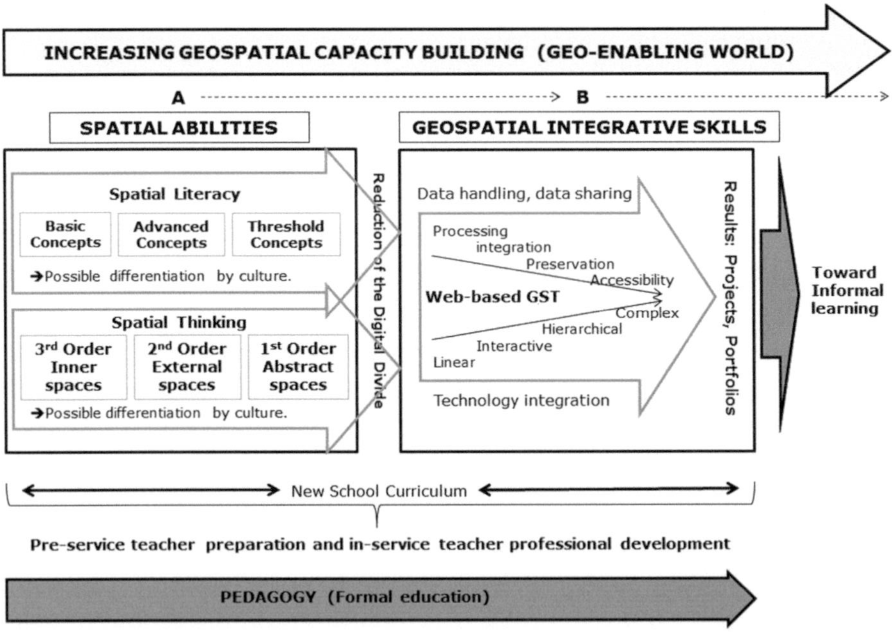

Fig. 17.1 The novel model as a basic guide for integral geography education with GST in a geo-enabling world (Muñiz-Solari 2014)

physical and human events or both at the same time, GST practitioners have to confront an increasing evidence of multi-dimensional data integration (MacEachren and Kraak 2001; Yang et al. 2010). Therefore, data processing, data integration, data preservation and accessibility constitute important conditions of the geospatial data life cycle. Integrative geospatial information also entails to study the problem of real time change that makes any digital map and its quantitative information obsolete in a short period of time. It is not only about diverse spatial transformation of any phenomenon but a continuous renovation through time.

Teachers and students learning in collaborative work must realize that their GP will depend on multi-dimensional data integration and real-time information. A new curriculum must include procedures for learning data handling in order to facilitate geospatial knowledge. Multi-dimensional data integration and data sharing will get more evident with Cloud computing and Service Oriented Architecture (SOA) techniques (El-Sharkawi et al. 2013; Doytsher 2013). Kerski in Chap. 15 coincides with this approach when he mentions Sofware-as-a-Service (SaaS) and Cloud. Data sharing is also an important component of the curriculum. Teachers and students must learn how to deal with data sharing. The National States Geographic Information Council strongly believes that open sharing of geospatial data is in the best interest of our communities (National States Geographic Information Council [NSGIC] 2011). Practitioners need to comply with laws that govern

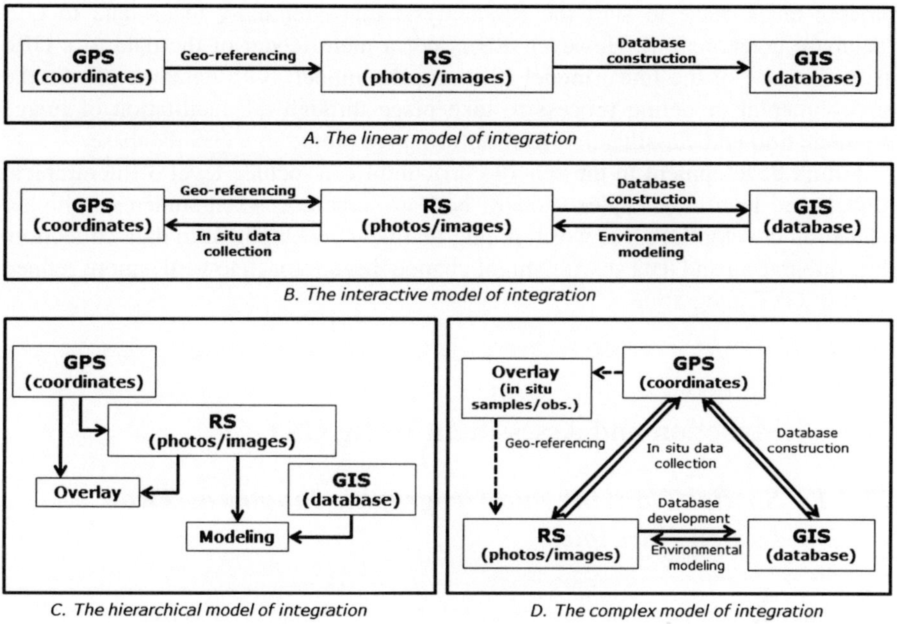

Fig. 17.2 GST models of integration (Adapted from Gao 2002)

how data can be used. In Europe, INSPIRE Directive initiative should be included in the geographical education probably in the context of GIS. Teaching materials also should be elaborated parallel to the development of metadata editor (Bartha and Kocsis 2011).

Besides data, cyberinfrastructure is a combination of network protocols, computing platforms, and computational services that brings practitioners, information, and computational tools together to perform science (Yang et al. 2010). GST, as geospatial cyberinfrastructure, function with different levels of integration within the Web-based environment. The current trend is the integration of different technologies (Aina 2012). According to Gao (2002), there are four possible levels of integration which are shown in Fig. 17.2: (1) Linear, (2) Interactive, (3) Hierarchical, and (4) Complex.

Level 1 (linear model) implies that the three components are not equally important. GIS plays a dominant role as final destination of this process. This model is the typical and most common in GP in secondary education, yet it is often disconnected from RS and GPS when data is not produced but given as part of GP. One typical example is minimal GIS as presented by Hong in Chap. 10 to learn GST while improving spatial thinking skills. We propose level 2 (interactive model) as the most adequate approach to build a curriculum for GP with GST in formal education in order to reach a reasonable integration. Data can flow mutually or directed and received in equal amount among the main GSTs. In other words, a given student team records data using GPS equipment, then the geo-referencing process takes place to start the RS analysis that eventually feeds data to GIS (database construction). However, RS is not a mere feeder of the data to a GIS, as in the case of the lineal model (the simplest model). GIS database allows the environmental modeling process to take place through the facilitation of image segmentation and classification from information stored in a GIS database.

Future development in the school curriculum can include level 3 (hierarchical model) and level 4 (complex model); both are common when students in higher education develop GP. It is worth noting that new development in data collection, data integration and data sharing might change this approach toward a more refined form of GST integration.

17.3 Instruction and Assessment Using GST

17.3.1 Spatial Thinking and Geographic Inquiry as Two Design Principles

The potential and possible benefits of using GST to support a variety of goals in education are well presented by Demirci in Chap. 12 and elsewhere (e.g., Goldstein and Alibrandi 2013; Sinton 2009; Tsou and Yanow 2010). Nevertheless, using GST can be neither a panacea for the achievement of all different types of learning goals nor a replacement for all the existing practices of instruction and assessment. For example, when instruction focuses on the acquisition and retention of declarative knowledge that is not necessarily spatial, GST may not be the most effective tool for attaining such knowledge (e.g., Liu et al. 2010). Research is insufficient to address questions and concerns about whether and how effective GST can be for teaching and learning specific knowledge, skills, and practices in geography. It is crucial that research carefully examines existing curricula and identifies the specific learning goals that GST can best help to accomplish.

Despite the uncertainty about the superior effects of GST and GP to other instructional approaches and technologies for teaching and learning, convincing evidence has emerged that GST can help to achieve at least two important goals in geography education: spatial thinking and geographic inquiry. GST offers teachers and students more possibilities to engage in spatial thinking and sophisticated geographic inquiry than would otherwise be possible (NRC 2006; Liu and Zhu 2008). For example, Geographic Information Systems (GIS) allows students to visualize spatial patterns from large, complex, multi-layered datasets quickly and easily, and allows alternative approaches to problems and perspectives to view phenomena from a spatial perspective. GIS can also facilitate spatial reasoning through 3D visualization, overlay analysis, and spatial pattern analysis (Cheung et al. 2011; Lee and Bednarz 2009; Perkins et al. 2010), which is the essence of spatial thinking practice. In addition, the power of GIS in education presents opportunities for students to ask questions of data and thus become able to initiate geographic inquiry processes. Using this inquiry-driven approach, students can learn to formulate research questions, develop research methodologies, collect and analyze data, and draw conclusions (Cheung et al. 2011). GIS is an effective tool for what-if analysis, as it supports functions for modelling the complexities of real-world problems. Using GIS, students can make geospatial inquiries to explore various real-world problems (Hwang 2013). Research suggests that extended use of GIS improves students' ability to learn methods of geographic inquiry, such as posing geographic questions, collecting relevant data, analyzing data, and finding and communicating potential answers (Kolvoord et al. 2011). We have proposed spatial abilities and geospatial integrative skills as two key components of the new model that would better support geospatial literacy. The geographic inquiry process may well mesh with the philosophy of teaching with GIS (Kerski 2008), and it not only requires but also facilitates the practice of geospatial integrative skills. Therefore, spatial thinking and geographic inquiry should serve as two key principles for the design of instruction and assessment of geospatial literacy.

However, neither GST nor GP themselves result in spatial thinking and inquiry-based learning. The merit of using GST in the classroom largely depends upon how appropriately they are used to facilitate students' practicing thinking skills and scientific inquiry. Research indicates that instruction using GIS produces a positive effect on spatial learning only when it emphasizes the understanding, not just the acquisition and retention, of concepts and the application of the knowledge learned (Bednarz 2004; Cheung et al. 2011). Teaching for understanding has little to do with traditional methods of teaching, and it requires students to play an active role in the learning process, as they must learn at a higher, more abstract level (Bednarz 2004). Therefore, to teach effectively with GST, teachers must develop a deeper knowledge of the subject matter and skills to implement GST with pedagogical strategies such as problem-based, team-based, and inquiry-driven learning (Bednarz and Van der Schee 2006). Instructional and assessment practices should be aligned with the key components of spatial thinking and geographic inquiry: students' understanding of geographic and spatial concepts, ability to use and create a variety of representations to communicate information, and practice of problem solving through spatial analysis and reasoning in an authentic context.

17.3.2 Challenges to Geospatial Instruction and Assessment

It is true, especially in the formal education system, that the teacher plays a role as a curricular-instructional gatekeeper (Thornton 2001). A fundamental role of teachers in the adoption of educational reform and innovation (Bednarz and Van der Schee 2006), including GST, cannot be over-emphasized. As Kerski points out in Chap. 15, instructional challenges, including resolving uncertainty inherent in evolving technologies, limited time and opportunities for teacher education, and inadequate resources for the best approaches to teaching with GST, are major constraints on the effective integration of GST and GP into instructional and assessment practices. Moreover, research has not yet provided much insight to help teachers design or employ GST-supported spatial thinking and inquiry lessons. Little attention has been paid to "issues related to pedagogy and GIS" (Bednarz 2004, p. 197). Research should "provide more insights on how to make the best use of GIS in geography teaching" (Lam et al. 2009, p. 72).

The lack of assessment instruments that can be used in the classroom is also a hindrance to the integration of geospatial literacy into teaching and learning practice. Kerski (2007, 2008) suggests that teaching with GIS opens the possibility that teachers can design authentic tasks and assessments for students in which they evaluate student performance based on the students' portfolios, presentations, or projects, rather than multiple-choice and true-false examinations. However, there are few assessment instruments available for teachers to adopt in order to evaluate their students' geospatial literacy. The Spatial Thinking Ability Test (STAT) (Lee and Bednarz 2009) is a well-known, rigorously developed standardized test of geospatial literacy, but as Sharpe and Huynh point out in Chap. 14, it has not been widely adopted into instructional practice, partly due to a lack of teacher awareness and relevant training.

Indeed, preparing teachers who are capable of designing and implementing such GST-supported inquiry-based projects and authentic assessment is a big challenge. It requires that the concepts of GST and GP be explicitly integrated into the curricula of teacher preparation programs, which necessitate a reform of teacher preparation curricula through close cooperation between geography and education faculties. A minimum condition to be met is ensuring geography coursework in teacher preparation. It is very likely that geographically illiterate teachers cannot conduct spatial analyses, let alone teach students to do them, despite feeling comfortable with the technological aspects of GIS and problem-based or inquiry-based learning. Research is needed to identify the characteristics of teacher preparation programs that have a positive impact on teacher knowledge, skills, and dispositions related to their use of GST for teaching.

Central to the increasing capacity building boosted by the geo-enabling world is the need for an effective reduction of the digital divide. Programs for pre-service teacher preparation and in-service teacher development must reduce the gap between teachers as digital immigrants and their students as digital natives. This is more evident with digital natives in formal education who are taught by digital immigrants (teachers), who have to learn and adapt to using emerging technologies (Prensky 2001). Yet the gap between students and their teachers is neither equal nor a chasm so large that it cannot be reduced. Tapscott (1999) already indicated that the attention should be turned to the 'Net generation'. After the revision of empirical research conducted in more than 20 countries across several continents, Jones and Shao (2011), studying the net generation and digital natives, point out that "universities need to pay more attention to aspects of the educational design of courses and programmes in addition to the design of teaching materials and resources" (p. 44). Haulset and Kelzer (2012) present evidence that "pre-service teachers may be better served by geospatial science education that is integrated into their undergraduate education" (p. 163).

17.3.3 Building Teacher Capacity for Successful Implementation

More research should be done to provide insights into the best use of GST and GP in geography teaching. Favier and Van der Schee's (2012) study is one of few examples that deal with this issue, providing teachers with insights into the design and implementation of GIS-supported inquiry-based projects. According to the authors, four characteristics represent a desirable GIS-supported inquiry-based project. First, it has clear and attainable learning goals that focus on a deeper understanding of geographic inquiry. Second, it is grounded in good domain-specific questions and hypotheses. Third, it is free from stuck-points while it provides opportunities to carry out the whole process of geographic inquiry, including data collection, analysis, and the visualization of the results using GIS. Finally, it offers a considerable amount of guidance to fully achieve as many learning goals as possible. A line of research such as this example will definitely contribute to the development of materials and resources for effective teaching with GST.

A good example of how to study the incorporation of Information Communication Technology (ICT) pedagogy in pre-service teachers is the Australian case. The Teaching Teachers for the Future (TTF) project, which is based on the construct of the Technological Pedagogical Content Knowledge (TPACK) framework (Koehler and Mishra 2009). The TTF project considers it very important that ICT be integrated and presented in various elements of teacher preparation programs (Albion et al. 2011). An important element of the TTF project is its auditing of teacher preparation programs. This process should be emulated if we want to determine how GST is integrated and presented in various elements of the curriculum.

We also propose that research on GST in geography education should pay more attention to the concept of teacher disposition. Teacher disposition can be defined as "the tendencies of a teacher's behavior employing particular knowledge and skills to achieve certain teaching goals" (Jo and Bednarz 2014, p. 199). Teacher education researchers have examined the concept of teacher disposition for over a decade, and they now agree that a teacher's disposition, in addition to their knowledge and skills, are an important factor in effective teaching (e.g., Dottin 2009; Lee and Hemer-Patnode 2010; Miller and Maninger 2012; National Council for the Accreditation of Teacher Education [NCATE] 2002, 2008).

Besides the level of knowledge and skills, experiences during teacher preparation substantially influence pre-service teachers' conceptualization of geography as a discipline. It is apparent that such an internal conceptualization influences their willingness to learn about and to adopt innovative approaches like GST to their instruction and assessment practice. As Sanchez (2009) stresses:

> The changes depend on the willingness of teachers to change their practices. It depends on their acceptance to change their roles in the classroom. It depends also on their agreement to being involved in open-ended projects in which they can face difficulties in solving

problems. These points are probably the critical dimensions to be taken into account to support the introduction of geotechnologies in secondary education (p. 71).

Likewise, a teacher's knowledge and skills of GST are necessary but not sufficient for effective teaching with GST, because they do not guarantee the teacher's inclination and willingness to employ them in the classroom. Understanding teachers' beliefs about and inclinations towards teaching with GST and how these beliefs and inclinations shape their implementation practices should be a priority for research in the near future.

Among the six models of in-service GST teacher training presented in this book (see Chap. 10) we suggest a community partnership model and a snowball dispersion model. Both models are designed to emphasize collaboration, which is a critical component for developing a community of practice approach when using GST. Collaboration also facilitates interdisciplinary work and international networking (Muñiz-Solari and Coats 2009). Furthermore, collaboration eases problem solving through project designs, which should be the main result of GST work and GP to engage teachers and students with local communities. One of the best examples of projects developed by teachers and students using mostly GPS and GIS are those presented by more than 30 countries (Milson et al. 2012). The importance of collaborative efforts is also stressed by Bryant and Favier (see Chap. 11), as they conceptualize Technological Pedagogical Content Knowledge (TPACK) for teaching with GIS and propose a collaborative inquiry model as a strategy for the development of TPACK.

Through collaborative work in formal education, each participant learns the process of being part of a community of practice, which is the first step to becoming integrated as a full citizen of the world. Citizenship does not work by policies and political rhetoric but through active collaboration. We propose that the practice of geography education in all schools must be performed by both teachers and students, as they constitute communities of practice. First, the learning environments have to be active in collaborative work to create neither a teacher-centered approach, nor a student-centered focus, but an integrative group of practitioners evolving to become teams. Second, the new curricula, which must be adapted to different cultural conditions, have to create a coordinated relationship between spatial abilities and geospatial integrative skills. Third, teachers acting as facilitators or guiders need to implement inquiry-based learning approaches.

Collaborative work in the school environment has paved the way to develop team work based on inquire-based learning and problem solving. By doing so, teachers and students initiate the process by building up 'communities of practice'. This term attributed to Lave and Wenger (1991) represents the integration of a group of practitioners who learn by constant transmission of knowledge built by their members. ICT and the Web 2.0 are facilitating the process of community of practice in the schools allowing their participants in geography education to develop GST and GP integrated with the worldwide community.

17.4 Conclusions

The need for spatial abilities to use GST requires spatial literacy which is built through several levels of conceptualization. We propose an agenda to activate the school curriculum that develops several orders of spatial thinking. Therefore, practices with GST should be related to spatial thinking *in* space, *about* space, and *with* space. As vital components of spatial thinking, cognitive processes should be organized in the cognitive domains (i.e., information gathering, building understanding, productive thinking) as the most important to work with GST and develop GP.

Geospatial knowledge should be facilitated by learning procedures that will ease data handling and data sharing. With the current trend of technology integration it seems advisable to encourage the learning process about GST and with GST with an interactive approach. The integration under that approach of technologies such as GPS, RS, and GIS will allow data to flow mutually among the main GSTs.

The call for global geospatial education has to be initiated by new curricula in the schools around the world that make the connection between GST and participatory community research. In that regard, formal education is not isolated from informal learning but a complement of a common objective that is the constant education for life. One good road to propose this linkage is the creation of curriculum structure based on real projects and portfolios of local and regional guides for geospatial problem resolution.

Research suggests that GST and GP facilitate learning to think spatially and geographic inquiry. Therefore, instructional and assessment practices using GST can be more effective when they focus on students' understanding of geographic and spatial concepts, ability to use and create a variety of representations to communicate information, and practice of problem solving through spatial analysis and reasoning in an authentic context. More research is needed to enlighten whether and how GST and GP can help accomplish other important educational goals than spatial thinking and geographic inquiry.

A big challenge to geospatial instruction and assessment is preparing teachers who are capable of designing and implementing GST-supported inquiry-based projects and authentic assessment. A close collaboration between geography and education faculties to integrate GST and GP into teacher preparation programs is critical. Empirical research on characteristics of teacher preparation programs that have a positive impact on teacher knowledge, skills, and dispositions related to their use of GSTs for teaching should shed light on this matter.

Ultimately, geography education researchers bear a critical and unavoidable obligation. Formal and informal geography education should be designed to foster geospatially literate generations that face technological and pedagogical challenges.

References

Aina, Y. A. (2012). Applications of geospatial technologies for practitioners: An emerging perspective of geospatial education. In S. J. Miah (Ed.), *Emerging informatics – Innovative concepts and applications* (pp. 3–20). Rijeka: InTech. http://cdn.intechopen.com/pdfs-wm/35692.pdf. Accessed 1 Oct 2014.

Albion, P., Jamieson-Proctor, R., & Finger, G. (2011). Age-related differences in ICT access and confidence among pre-service teachers. In G. Williams, N. Brown, M. Pittard, & B. Cleland (Eds.), *ASCILITE 2011 conference: Changing demands, changing directions* (pp. 21–32). Wrest Point-Hobart: ASCILITE

Bartha, G., & Kocsis, S. (2011). Standardization of geographic data: The European inspire directive. *European Journal of Geography, 2*(2), 79–89.

Bednarz, S. W. (2004). Geographic information systems: A tool to support geography and environmental education? *GeoJournal, 60*(2), 191–199.

Bednarz, S. W., & Van der Schee, J. A. (2006). Europe and the United States: The implementation of geographic information systems in secondary education in two contexts. *Technology, Pedagogy and Education, 15*(2), 191–205.

Cheung, Y., Pang, M., Lin, H., & Lee, C. K. J. (2011). Enable spatial thinking using GIS and satellite remote sensing: A teacher-friendly approach. *Procedia Social and Behavioral Sciences, 21*, 130–138.

Cousin, G. (2010). Neither teacher-centred nor student-centred: Threshold concepts and research partnership. *Journal of Learning Development in Higher Education, 2*, 1–9.

Demirci, A., Karaburun, A., Ünlü, M., & Özey, R. (2011). Using GIS-based projects in learning: Students help disabled pedestrians in their school district. *European Journal of Geography, 2* (2), 48–61.

De Vecchis, G., Pasquinelli D'Allegra, D., & Pesaresi, C. (2011). Geography in Italian schools: An example of a cross-curricular project using geospatial technologies for a practical contribution to educators. *Review of International Geographical Education Online, 1*(1), 4–25.

Dottin, E. S. (2009). Professional judgment and dispositions in teacher education. *Teaching and Teacher Education, 25*(1), 83–88.

Doytsher, Y. (2013). Innovations in the geospatial data technologies. TS 1A – The importance of geospatial information. In *The Proceedings of FIG Pacific Small Island developing states symposium 2013: Policies and practices for responsible governance*. Suva: International Federation of Surveyors. http://www.fig.net/pub/fiji/papers/ts%201a_doytsher_6736.pdf. Accessed 7 Oct 2014.

El-Sharkawi, A., Shouman, A., & Lasheen, S. (2013). Service oriented architecture for remote sensing satellite telemetry data implemented on Cloud computing. *I.J. Information Technology and Computer Science, MECS, 7*, 12–26.

Favier, T. T., & Van der Schee, J. A. (2012). Exploring the characteristics of an optimal design for inquiry-based geography education with geographic information systems. *Computers & Education, 58*(1), 666–677.

Gao, J. (2002). Integration of GPS with remote sensing and GIS: Reality and prospect. *Photogrammetric Engineering & Remote Sensing, 68*(5), 447–453.

Goldstein, D., & Alibrandi, M. (2013). Integrating GIS in the middle school curriculum: Impacts on diverse students' standardized test scores. *Journal of Geography, 112*(2), 68–74.

Golledge, R. G., Marsh, M., & Battersby, S. (2008). Matching geospatial concepts with geographic educational needs. *Geographical Research, 46*(1), 85–98.

Hagevik, R. (2011). Five steps to success: Implementing geospatial technologies in the science classroom. *Journal of Curriculum and Instruction, 5*(1), 34–53.

Hauselt, P., & Helzer, J. (2012). Integration of geospatial science in teacher education. *Journal of Geography, 111*(5), 163–172.

Hertzog, C. J., & Lieble, C. (1996). A study of two techniques for teaching introductory geography: Traditional approach versus cooperative learning in the university classroom. *Journal of Geography, 9*, 274–280.

Huynh, N. T., & Dean, N. (2011). GIS instruction: Learning from student perception and concept difficulty. *Research in Geographic Education, 13*(2), 52–78.

Hwang, S. (2013). Placing GIS in sustainability education. *Journal of Geography in Higher Education, 37*(2), 276–291.

Jo, I., & Bednarz, S. W. (2014). Dispositions toward teaching spatial thinking through geography: Conceptualization and an exemplar assessment. *Journal of Geography, 113*(5), 198–207.

Jones, C., & Shao, B. (2011). *The net generation and digital natives: Implications for higher education.* York: Higher Education Academy. http://oro.open.ac.uk/30014/1/Jones. Accessed 2 Oct 2014.

Kerski, J. J. (2007). Geographic information systems in education. In J. P. Wilson & A. S. Fotheringham (Eds.), *The handbook of geographic information science* (pp. 540–556). Oxford: Blackwell Publishing Ltd.

Kerski, J. J. (2008). The role of GIS in Digital Earth education. *International Journal of Digital Earth, 1*(4), 326–346.

Kerski, J. J., Milson, A. J., & Demirci, A. (2012). Synthesis: The future landscape of GIS in secondary education. In A. J. Milson, A. Demirci, & J. J. Kerski (Eds.), *International perspectives on teaching and learning with GIS in secondary schools* (pp. 315–326). Dordrecht: Springer.

Knauff, M. (2013). *Space to reason.* Cambridge, MA: MIT Press.

Koehler, M. J., & Mishra, P. (2009). What is technological pedagogical content knowledge? *Contemporary Issues in Technology and Teacher Education, 9*(1), 60–70.

Kolvoord, R. A., Uttal, D. H., & Meadow, N. G. (2011). Using video case studies to assess the impact of the use of GIS on secondary students' spatial thinking skills. *Procedia Social and Behavioral Sciences, 21*, 372–379.

Lam, C. C., Lai, E., & Wong, J. (2009). Implementation of geographic information system (GIS) in secondary geography curriculum in Hong Kong: Current situations and future directions. *International Research in Geographical and Environmental Education, 18*(1), 57–74.

Lave, J., & Wenger, E. (1991). *Situated learning: Legitimate peripheral participation.* New York: Cambridge University Press.

Lee, J., & Bednarz, R. S. (2009). Effect of GIS learning on spatial thinking. *Journal of Geography in Higher Education, 33*(2), 183–198.

Lee, Y. A., & Hemer-Patnode, L. (2010). Developing teacher candidates' knowledge, skills, and dispositions to teach diverse students. *Journal of Instructional Psychology, 37*(3), 222–235.

Lisec, A., & Ruiz Fernandez, L.A. (2008). European Project on higher education in the fields related to geomatics as support for mobility of students and teachers. *ISPRA-The International Archives of the Photogrammetry, Remote Sensing and Spatial Information Sciences. Vol. XXXVII. Part B6a.* Beijing, China.

Liu, S., & Zhu, X. (2008). Designing a structured and interactive learning environment based on GIS for secondary geography education. *Journal of Geography, 107*(2), 12–19.

Liu, Y., Bui, E. N., Chang, C. H., & Lossman, H. G. (2010). PBL-GIS in secondary geography education: Does it result in higher-order learning outcomes? *Journal of Geography, 109*(4), 150–158.

Livingstone, D., & Lynch, K. (2000). Group project work and student-centred active learning: Two different experiences. *Studies in Higher Education, 25*(3), 325–345.

MacEachren, A. M., & Kraak, M. J. (2001). Research challenges in geovisualization. *Cartography and Geographic Information Science, 28*(1), 3–12.

Meyer, J., & Land, R. (2005). Threshold concepts and troublesome knowledge (2): Epistemological considerations and a conceptual framework for teaching and learning. *Higher Education, 49*, 373–388.

Miller, M. S., & Maninger, R. M. (2012). Teachers' dispositions: One collegiate journey. *National Teacher Education Journal, 5*(4), 101–106.
Milson, A. J., Demirci, A., & Kerski, J. J. (Eds.). (2012). *International perspectives on teaching and learning with GIS in secondary schools*. New York: Springer.
Misheck, M., Ezra, P., & Mandoga, E. (2013). Geographic literacy and world knowledge amongst open distance learning students in Zimbabwe. *Greener Journal of Educational Research, 3*(7), 301–309.
Mishra, P., & Koehler, M. J. (2006). Technological pedagogical content knowledge: A new framework for teacher knowledge. *Teachers College Record, 108*, 1017–1054.
Morseley, D. M. J., Gregson, E. M., & Higgins, S. (2005). Thinking skills frameworks for use in education and training. *British Educational Research Journal, 31*(3), 367–390.
Muñiz-Solari, O. (2014, October 16). *Geospatial technologies in geography education*. East China Normal University. Shanghai. School of Geographical Science. Colloquium.
Muñiz-Solari, O., & Coats, C. (2009). Integrated network: National and international online experiences. *International Review of Research in Open and Distance Learning, 10*(1), 1–19.
National Council for the Accreditation of Teacher Education. (2002). *Professional standards for the accreditation of schools, colleges, and departments of education*. Washington, DC: NCATE.
National Council for the Accreditation of Teacher Education. (2008). *Professional standards for the accreditation of teacher preparation institutions*. Washington, DC: NCATE.
National Research Council. (2006). *Learning to think spatially: GIS as a support system in the K-12 curriculum*. Washington, DC: National Academy Press.
National Research Council. (2010). *Understanding the changing planet*. Washington, DC: The National Academies Press.
National States Geographic Information Council. (2011). *Geospatial data sharing. Guidelines for best practices*. NSGIC. http://www.nsgic.org/public_resources/NSGIC_Data_Sharing_Guidelines_120211_Final.pdf. Accessed 1 Oct 2014.
Newcombe, N. S. (2000). So, at last we can begin. *Developmental Science, 3*(3), 276–278.
Newcombe, N. S., & Shipley, T. F. (2014). Thinking about spatial thinking: New typology, new assessments. In J. S. Gero (Ed.), *Studying visual and spatial reasoning for design creativity* (pp. 1–13). New York: Springer.
O'Reilly, T. (2005). *What is Web 2.0: Design patterns and business models for the next generation of software*. http://oreilly.com/web2/archive/what-is-web-20.html. Accessed 1 Oct 2014.
Oyserman, D., & Lee, S. (2008). Does culture influence what and how we think? Effects of priming individualism and collectivism. *Psychological Bulletin, 2*, 311–342.
Pawson, E., Fournier, E., Haigh, M., Muniz, O., Trafford, J., & Vajoczki, S. (2006). Problem-based learning in geography: Towards a critical assessment of its purposes, benefits and risks. *Journal of Geography in Higher Education, 30*(1), 103–116.
Perkins, N., Hazelton, E., Erickson, J., & Allan, W. (2010). Place-based education and geographic information systems: Enhancing the spatial awareness of middle school students in Maine. *Journal of Geography, 109*(5), 213–218.
Prensky, M. (2001). Digital natives, digital immigrants. *On the Horizon, 9*(5), 1–6.
Sanchez, E. (2009). Innovative teaching/learning with geotechnologies in secondary education. In A. Tatnall & A. Jones (Eds.), *9th IFIP TC 3 world conference on computers in education: Education and technology for a better world* (pp. 65–74). Berlin: Springer.
Schneider-Sliwa, R., & Leser, H. (2003). Geography – Problem solving competencies for societal concerns. *Geographica Helvetica, 58*(3), 176–183.
Schoeller Sprague, D., & Boyon, N. (2011). National Geographic Society-Roper 2002. Global geographic literacy survey. In G. Elbow, D. Rutherford, & C. Shearer (Eds.), *Geographic literacy in the United States: Challenges and opportunities in the NCLB era* (pp. 6–10). Washington, DC: NCGE.

Schyvens, R., Griffin, A. L., Jocoy, C. L., Liu, Y., & Bradford, M. (2008). Experimenting with active learning in geography: Dispelling the myths that perpetuate resistance. *Journal of Geography in Higher Education, 32*(1), 51–69.

Sinton, D. S. (2009). Roles for GIS within higher education. *Journal of Geography in Higher Education, 33*(1), S7–S16.

Srivastava, S. K. (2012). Threshold concepts framework: An effective approach to address future spatial sciences pedagogical challenges. In *The Proceedings of The Queensland surveying and spatial conference 2012: The future of surveying and spatial science is open*. Brisbane, Australia. http://www.sssi.org.au/userfiles/docs/QLD%20Region/documents_1349830710.pdf. Accessed 7 Oct 2014.

Sui, D., & DeLyser, D. (2012). Crossing the qualitative-quantitative chasm I: Hybrid geographies, the spatial urn, and volunteered geographic information (VGI). *Progress in Human Geography, 31*(1), 111–124.

Tang, G., Dong, Y., Jia, Y., & Gao, Y. (2008). Geo-spatial information technology education in China, present and future. *ISPRA-The International Archives of the Photogrammetry, Remote Sensing and Spatial Information Sciences. Vol. XXXVII. Part B6a*. Beijing, China.

Tapscott, D. (1999). Education the net generation. *Educational Leadership, 56*(5), 6–11.

Tejpal, Sharma, M., & Jaglan, M. S. (2012). Significance of geospatial technology and status of its education and training in India. *International Journal of Geomatics and Geosciences, 2*(3), 730–737.

Thornton, S. (2001). Educating the educators: Rethinking subject matter and methods. *Theory into Practice, 40*(1), 72–79.

Timmermans, J. A. (2010). Changing our minds. The developmental potential of threshold concepts. In J. H. F. Meyer, R. Land, & C. Baillie (Eds.), *Threshold concepts and transformational learning* (pp. 3–19). Rotterdam: Sense Publishers.

Tsou, M. H., & Yanow, K. (2010). Enhancing general education with geographic information science and spatial literacy. *URISA Journal, 22*(2), 45–54.

Tuna, F. (2012). Student's perspectives on active learning in geography: A case study of level of interest and usage in Turkey. *European Journal of Educational Studies, 4*(2), 163–175.

Tversky, B. (2005). Functional significance of visuospatial representations. In P. Shah & A. Miyake (Eds.), *Handbook of higher-level visuospatial thinking*. Cambridge: Cambridge University Press.

Vyas, A., & Koening, G. (2014). Computer aided teaching in photogrammetry, remote sensing, and geomatics – A status review. ISPRS Technical Commission VI Symposium, Wuhan, China. *The International Archives of the Photogrammetry, Remote Sensing and Spatial Information Sciences, XL-6*.

Yang, C., Raskin, R., Goodchild, M., & Gahegan, M. (2010). Geospatial cyberinfrastructure: Past, present and future. *Computers, Environment and Urban Systems, 34*, 264–277.

MIX
Papier aus verantwortungsvollen Quellen
Paper from responsible sources
FSC® C105338

If you have any concerns about our products,
you can contact us on
ProductSafety@springernature.com

In case Publisher is established outside the EU,
the EU authorized representative is:
Springer Nature Customer Service Center GmbH
Europaplatz 3, 69115 Heidelberg, Germany

Printed by Libri Plureos GmbH
in Hamburg, Germany